AF726549

Textbook of Geology

Textbook of Geology

Collin Shephard

MURPHY & MOORE
www.murphy-moorepublishing.com

Published by Murphy & Moore Publishing,
1 Rockefeller Plaza,
New York City, NY 10020, USA
www.murphy-moorepublishing.com

Textbook of Geology
Collin Shephard

© 2022 Murphy & Moore Publishing

International Standard Book Number: 978-1-63987-520-7 (Hardback)

This book contains information obtained from authentic and highly regarded sources. All chapters are published with permission under the Creative Commons Attribution Share Alike License or equivalent. A wide variety of references are listed. Permissions and sources are indicated; for detailed attributions, please refer to the permissions page. Reasonable efforts have been made to publish reliable data and information, but the authors, editors and publisher cannot assume any responsibility for the validity of all materials or the consequences of their use.

Trademark Notice: Registered trademark of products or corporate names are used only for explanation and identification without intent to infringe.

Cataloging-in-Publication Data

Textbook of geology / Collin Shephard.
p. cm.
Includes bibliographical references and index.
ISBN 978-1-63987-520-7
1. Geology. 2. Earth sciences. I. Shephard, Collin.
QE26.3 .G46 2022
550--dc23

TABLE OF CONTENTS

PREFACE

The branch of Earth science which is concerned with the study of materials which make up the Earth is known as geology. It attempts to describe the structure of the Earth and the processes which helped in shaping its structure. The subject of geology can be classified into petrology, structural geology, planetary geology, economic geology and engineering geology. One of the important aspects of this discipline is to determine the age of rocks. This is achieved by employing the principles of intrusive relationships, cross-cutting relationships, superposition, faunal succession and original horizontality. The dating systems studied under geology include radiocarbon dating, uranium-lead dating and argon-argon dating. Optically stimulated luminescence and cosmogenic radionuclide are some of the recent techniques used for dating surfaces and erosion rates. This book is a compilation of chapters that discuss the most vital concepts in the field of geology. The topics included herein on geology are of utmost significance and bound to provide incredible insights to readers. This book will serve as a reference to a broad spectrum of readers.

Given below is the chapter wise description of the book:

Chapter 1- Geology is the study of the solid features of the Earth; the composition, landforms and history of the solid Earth. It also involves the study of the structure of Earth on and beneath its surface and the processes involved in shaping that structure. This is an introductory chapter which will introduce briefly all the significant aspects of Geology.

Chapter 2- Petrology focuses on the rocks and the conditions under which they form. Petrology has three major types: igneous, metamorphic and sedimentary petrology. Igneous petrology is the study of physical and chemical properties of igneous rocks while metamorphic petrology is the study of composition and texture of metamorphic rocks. All the diverse principles of petrology and its types have been carefully analyzed in this chapter.

Chapter 3- The branch of science that encompasses the scientific study of the distribution, movement and management of groundwater is known as hydrogeology. Engineering geology is an interdisciplinary field that involves the application of geology to engineering particularly geological and geotechnical engineering. This chapter will help the readers in understanding the basic aspects of hydrogeology and engineering geology.

Chapter 4- Structural geology is the study of rock deformation on a small scale as well as a large scale. The small scale study of the features of rocks can be done by mounting the ground rocks on glass slide and examining them with a polarizing microscope. This chapter has been carefully written to provide an easy understanding of the varied facets of structural geology.

Chapter 5- The scientific discipline that deals with the study of composition and texture of sedimentary rocks and the processes like transportation, deposition and lithification that are involved in the formation of sedimentary rocks is known as sedimentology. The aim of this chapter is to explore the topics related to sedimentology which are crucial for a complete understanding of the subject.

Indeed, my job was extremely crucial and challenging as I had to ensure that every chapter is informative and structured in a student-friendly manner. I am thankful for the support provided by my family and colleagues during the completion of this book.

Collin Shephard

Chapter 1

Geology: An Introduction

Geology is the study of the solid features of the Earth; the composition, land-forms and history of the solid Earth. It also involves the study of the structure of Earth on and beneath its surface and the processes involved in shaping that structure. This is an introductory chapter which will introduce briefly all the significant aspects of Geology.

Geology is a science, meaning that geological questions are investigated with deductive reasoning and scientific methodology. Geology is arguably the most interdisciplinary of all of the sciences because geologists must understand and apply other sciences, including physics, chemistry, biology, mathematics, astronomy, and more.

An aspect of geology that is unlike most of the other sciences is the role played by time deep time billions of years of it. When geologists study the evidence around them, they are often observing the results of are observing the results of events that took place thousands, millions, and even billions of years in the past, and which may still be ongoing. Many geological processes happen at incredibly slow rates millimetres per year to centimetres per year but because of the amount of time available, tiny changes can result in expansive oceans forming, or entire mountain ranges being worn away.

Geology is about understanding the evolution of Earth through time. It is about discovering resources such as metals and energy, and minimizing the environmental implications of our use of resources. It is about learning to mitigate the hazards of earthquakes, volcanic eruptions, and slope failures.

Why Study Earth?

Why? Because Earth is our home our only home for the foreseeable future and in order to ensure that it continues to be a great place to live, we need to understand how it works. Another answer is that some of us can't help but study it because it's fascinating. But there is more to it than that.

- We can study rocks and the fossils they contain to understand the evolution of our environment and the life within it.
- We can learn to minimize our risks from earthquakes, volcanoes, slope failures, and damaging storms.

- We can learn how and why Earth's climate changed in the past, and use that knowledge to understand both natural and human-caused climate change.
- We rely on Earth for resources such as soil, water, metals, industrial minerals, and energy, and we need to know how to find these resources and exploit them sustainably.
- We can recognize how human activities have altered the environment, and learn how to prevent and sometimes repair the damage.
- We can use knowledge of Earth to understand other planets in our solar system, and those around distant stars.

The Importance of Geological Studies for Minimizing Risks to the Public

Figure shows a slope failure that took place in January 2005 in the Riverside Drive area of North Vancouver. The steep bank beneath the house shown gave way, and a slurry of mud and sand flowed down. It destroyed another house below, and killed one person. The slope failure happened after a heavy rainfall, which is a common occurrence in southwestern B.C. in the winter.

Aftermath of a deadly debris flow in the Riverside Drive area of North Vancouver in January, 2005.

A geological report written in 1980 warned the District of North Vancouver that the area was prone to slope failure, and that steps should be taken to minimize the risk to residents. Unfortunately, not enough was done in the intervening 25 years to prevent a tragedy.

What do Geologists do?

Geologists do a lot of different things. Many of the jobs are the things you would expect. Geologists work in the resource industry, including mineral exploration and mining, and exploring for and extracting sources of energy. They do hazard assessment and mitigation (e.g., assessment of risks from slope failures, earthquakes, and volcanic eruptions). They study the nature of the subsurface for construction projects such as highways, tunnels, and bridges. They use information about the subsurface for water supply planning, development, and management; and to decide how best to contain contaminants from waste.

Geologists also do the research that makes practical applications of geology possible. Some geologists spend their summers trekking through the wilderness to make maps of the rocks in a particular location, and collect clues about the geological processes that occurred there. Some geologists work in laboratories analyzing the chemical and physical properties of rocks to understand how the rocks will behave when forces act on them, or when water flows through them. Some geologists specialize in inventing ways to use complex instruments to make these measurements. Geologists study fossils to understand ancient animals and environments, and go to extreme environments to understand how life might have originated on Earth. Some geologists help NASA understand the data they receive from objects in space.

Geological work can be done indoors in offices and labs, but some people are attracted to geology because they like to be outdoors. Many geological opportunities involve fieldwork in places that are as amazing to see as they are interesting to study. Sometimes these are locations where few people have ever set foot, and where few ever will again.

Geologists at work on the island of Spitsbergen, part of the Svalbard archipelago. The islands are located in the Arctic Ocean north of Norway.

Study of Earth using the Scientific Method

There is no single method of inquiry that is specifically the scientific method. Furthermore, scientific inquiry is not necessarily different from serious research in other disciplines. The key features of serious inquiry are the following:

- Creation of a hypothesis: This is a tentative idea about how to explain a set of observations.
- Test the hypothesis: A hypothesis can be used to make predictions, and experiments can be run to see if those predictions are correct.

An Example of the Scientific Method at Work

Consider a field trip to the stream shown in figure. Notice that the rocks in and along the stream are rounded off rather than having sharp edges. We might hypothesize that the rocks were rounded because as the stream carried them, they crashed into each other and pieces broke off.

Hypothesizing about the origin of round rocks in a stream.

If the hypothesis is correct, then the further we go downstream, the rounder and smaller the rocks should be. Going upstream we should find that the rocks are more angular and larger. If we were patient we could also test the hypothesis by marking specific rocks and then checking back to see if those rocks have become smaller and more rounded as they moved downstream.

If the predictions turn out to be correct, we must still be careful about how much certainty to attach to our hypothesis. Although our hypothesis might seem to us to be the only reasonable explanation, someone could argue that we have the mechanism wrong, and the rocks weren't rounded by bumping into each other. If our experiment didn't specifically check for the mechanism (e.g., by looking to see if chips fall off the rocks and the rocks are made smoother) then we would have to acknowledge the possibility. We needn't abandon the hypothesis as a useful tool for making predictions, but it is necessary to be open to the possibility that other things might be going on. If someone demonstrates conclusively that our hypothesis is wrong, then we have to discard the hypothesis and come up with a better one.

A good hypothesis is testable. Someone might argue that an extraterrestrial organization creates rounded rocks and places them in streams when nobody is looking. There is no practical way to test this hypothesis to confirm it, and there is no way to prove it false. Even if we never see aliens at work, we still can't say they haven't been, because according to the hypothesis they only work when people aren't looking. Compare this to the original hypothesis that allows us to make testable predictions, such as rocks getting smaller and rounder downstream. Our original hypothesis gives us a way to see how realistic it is, whereas the alien hypothesis gives us no way to know if it makes sense or not.

Theories and Laws

Two other terms appear in discussions of the scientific method: theory and law. A theory starts out as a hypothesis, but over a long period of time and a great many tests, it has never come up short. That doesn't mean it never will, but the odds of that are very unlikely given our present (and conceivable future) state of knowledge. You may have heard someone dismiss an idea by saying it is "just a theory," but they are using the term incorrectly if they mean to say it's a wild and unproven guess.

A law is a description of a phenomenon rather than an explanation of it. For example, you could do thousands of tests by dropping an object with known mass and measuring its acceleration and the force with which it hits the ground. Again and again your results will yield the formula force = mass × acceleration. However, that doesn't mean you know what is responsible for the force accelerating it toward the ground. Yes, we say that gravity is pulling it toward the Earth's surface, but why? A law is true regardless of why a phenomenon happens as long as it describes the outcome of that phenomenon. In geology there are three big ideas that are fundamental to the way we think about how Earth works. The ideas are like the sound track to a movie- sometimes we might not even notice them, but at the same time they affect our perception of what is happening.

Three Big Ideas: Geological Time, Uniformitarianism, and Plate Tectonics

Geological Time

Earth is approximately 4.57 billion years old (4,570,000,000 years), which is a long time for geological events to unfold and changes to happen. The changes themselves might be tiny. For example, over a year, a chemical reaction might eat away a few layers of atoms at the surface of a rock. But over time the changes accumulate and have a great impact. Over hundreds of millions of years the chemical reaction could cause a mountain range to crumble into grains of sand, and be swept away by rivers.

For geologists who study very, very slow processes, 10 million years might be a short time, and 1 million years might be trivial. For these geologists, intervals of 1 million years aren't even useful to consider, because the changes over that time are too small to see in the rocks that accumulated.

Expressing Geological Time in Numbers

Special notation is used for geological time because, as you might imagine, writing all those zeroes can become tiresome.

Table: Abbreviations used to describe geological time.

Abbreviation	Meaning	Example
Ga	giga annum or billions of years	Earth is 4.57 Ga old.
Ma	mega annum or millions of years	Earth is 4,570 Ma old.
ka	kilo annum or thousands of years	The last glacial cycle ended 11,700 years ago, or 11.7 ka.

Expressing Geological Time using the Geological Time Scale

The geological time scale is a way of breaking down geological time according to important events in Earth's history. Time is divided into eons, eras, periods, and epochs, and these intervals are referred to by names rather than by years. Giving intervals of geologic time names rather than using numbers makes sense because we won't always know the age in years (the absolute age) of a rock or fossil, but we can place it in context based on our knowledge of the geological record. We can describe its relative age by saying that it is older than or younger than another rock or fossil.

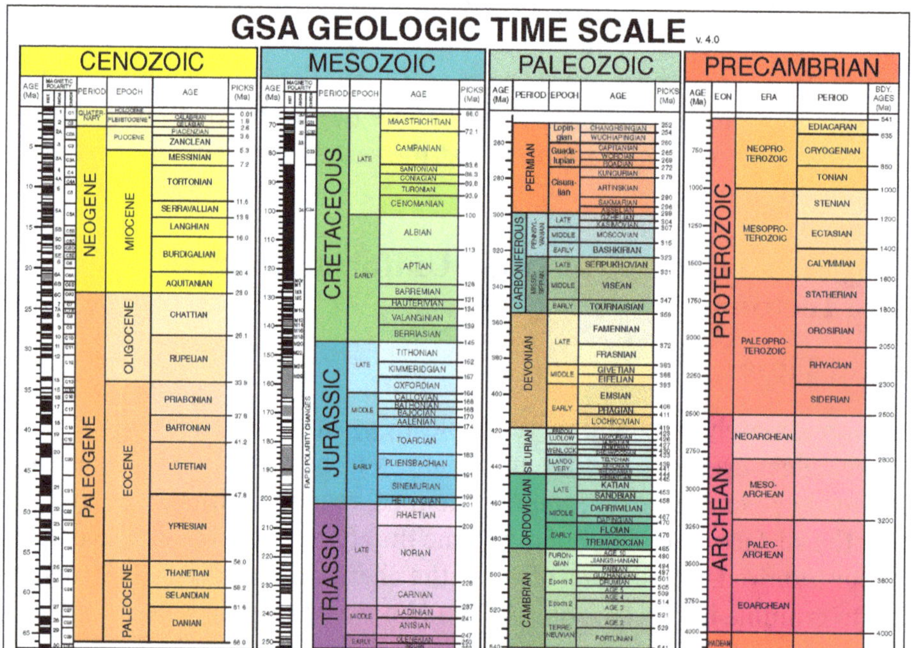

Geologic Society of America Geologic Time Scale, 2012.

The tricky thing about the geologic time scale is that the boundaries are always changing. As our knowledge of the absolute age of an event improves with new discoveries, it might be necessary to nudge a boundary earlier or later. Sometimes the original reason for defining a boundary no longer holds, but we agree to use it anyway. For example, the Phanerozoic Eon (the last 542 million years) is named for the time during which visible (phaneros) life (zoi) is present in the geological record, and its start was meant to mark the first appearance of these organisms. In fact, we now have evidence that large organisms those that leave fossils visible to the naked eye have existed longer than that, first appearing by 600 Ma at the latest.

An Early Definition of the Proterozoic

Notice that in figure the Proterozoic Eon precedes the Phanerozoic Eon. However, this was not always the case. Figure shows an excerpt from a periodical published in 1879, in which the Proterozoic is defined as covering the Cambrian through Silurian. The author refers to "the most extreme adherents of the Murchisonian party in geology," a reference to the contentious assertion by Scottish geologist Roderick Murchison that the Silurian Period should encompass the Cambrian and Ordovician periods as well.

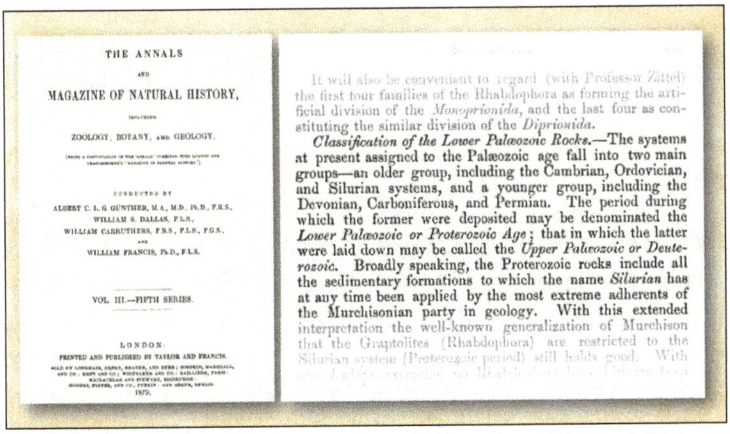

THE ANNALS AND MAGAZINE OF NATURAL HISTORY, INCLUDING ZOOLOGY, BOTANY, AND GEOLOGY.

CONDUCTED BY ALBERT C. L. G. GÜNTHER, M.A., M.D., Ph.D., F.R.S., WILLIAM S. DALLAS, F.L.S., WILLIAM CARRUTHERS, F.R.S., F.L.S., F.G.S., AND WILLIAM FRANCIS, Ph.D., F.L.S.

VOL. III.—FIFTH SERIES.

LONDON: PRINTED AND PUBLISHED BY TAYLOR AND FRANCIS. 1879.

It will also be convenient to regard (with Professor Zittel) the first four families of the Rhabdophora as forming the artificial division of the *Monoprionida*, and the last four as constituting the similar division of the *Diprionida*.

Classification of the Lower Palæozoic Rocks.—The systems at present assigned to the Palæozoic age fall into two main groups—an older group, including the Cambrian, Ordovician, and Silurian systems, and a younger group, including the Devonian, Carboniferous, and Permian. The period during which the former were deposited may be denominated the *Lower Palæozoic or Proterozoic Age*; that in which the latter were laid down may be called the *Upper Palæozoic or Deuterozoic*. Broadly speaking, the Proterozoic rocks include all the sedimentary formations to which the name *Silurian* has at any time been applied by the most extreme adherents of the Murchisonian party in geology. With this extended interpretation the well-known generalization of Murchison that the Graptolites (Rhabdophora) are restricted to the Silurian system (Proterozoic period) still holds good. With

An excerpt from the periodical The Annals and Magazine of Natural History (1879) in which the name "Proterozoic" is assigned to the Cambrian, Ordovician, and Silurian periods instead of to the time preceding the Cambrian.

A Way to think about Geological Time

A useful mechanism for understanding geological time is to scale it down into one year. The origin of the solar system and Earth at 4.57 Ga would be represented by January 1, and the present year would be represented by the last tiny fraction of a second on New Year's Eve. At this scale, each day of the year represents 12.5 million years; each hour represents about 500,000 years; each minute represents 8,694 years; and each second represents 145 years. Some significant events in Earth's history, as expressed on this time scale, are summarized in table.

Table: Some important dates expressed as if all of geological time were condensed into one year.

Event	Approximate Date	Calendar Equivalent
Formation of oceans and continents	4.5 - 4.4 Ga	First week of January
Evolution of the first primitive life forms	3.8 Ga	End of February
Formation of Saskatchewan's oldest rocks 3	3.4 Ga	End of March

Evolution of the first multi-celled animals	600 Ma	Beginning of November
Animals first crawled onto land	360 Ma	End of November
Vancouver Island reached North America and the Rocky Mountains were formed	90 Ma	December 16
Extinction of the non-avian dinosaurs	65 Ma	December 18
Beginning of the Pleistocene ice age	2 Ma	10:10 p.m., december 31
Oldest radiocarbon date from people living in Canada (British Columbia)	13.8 ka	11:58 p.m., december 31
Earliest evidence of human activity in Saskatchewan	11.5 ka	48 seconds before midnight, December 31
The last of the glacial ice retreats from Saskatchewan	6 ka	41 seconds before midnight, December 31
Hudson's Bay Company establishes a permanent settlement at Cumberland House in northern Saskatchewan	243 years ago	2 seconds before midnight, December 31

Uniformitarianism

Uniformitarianism is the notion that the geological processes occurring on Earth today are the same ones that occurred in the past. This is an important idea because it means that observations we make today about geological processes can be used to interpret and understand the rock record. While this idea might not seem remarkable today, it was ground breaking and even controversial for its time. Many people who heard about it for the first time thought about the age of the Earth in thousands of years, but uniformitarianism required them to think on timescales almost too vast to comprehend. For some, this implied questioning their most deeply held religious beliefs.

The Scottish geologist James Hutton initially presented the idea in 1785. Charles Lyell, also a Scottish geologist, paraphrased this idea as "the present is the key to the past" in his book Principles of Geology. This is how it is often described today.

To be clear, "the present is the key to the past" can be viewed as an oversimplification. Not all geological processes occurring today occurred at all times in the geological past. For example, some important chemical reactions that happened on Earth's surface today require abundant oxygen in the atmosphere, and could not have occurred prior to Earth developing an oxygen-rich atmosphere. Conversely, there was a time in Earth's history when continents as we know them hadn't yet developed. Some events, such as devastating impacts by objects from space, have never been witnessed on the same scale by humans. We must be cognizant of the fact that conditions were different at different times in Earth's history, and take that into account when interpreting the rock record.

Despite the different past conditions on Earth as a whole, there still exist environments today where some of these conditions are present. These environments are like little samples of what Earth used to be like. This means we can still use present conditions to inform us about the past, but we have to think carefully about ways that such environments today differ from the ancient environments that no longer exist.

Plate Tectonics

It is only within the last 50 years or so that we have been able to answer questions like, "How did that mountain range get there?" and "Why do earthquakes happen where they do?" The theory of plate tectonics- the idea that Earth's surface is broken into large moving fragments, called plates- profoundly changed our perspective on how the Earth works. Figure shows Earth's 15 largest tectonic plates, along with arrows indicating the plates' direction of motion, and how fast they go. (Longer arrows mean faster motion.) There are many more plates on Earth that are too small to show conveniently in figure.

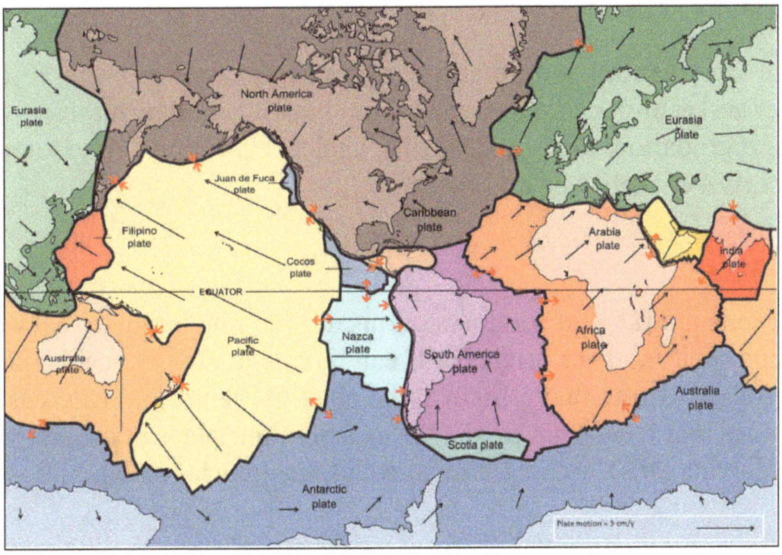

Earth's fifteen largest tectonic plates. Black arrows show the direction of plate motions. The length of the arrow indicates velocity. Red arrows show how plates move relative to each other.

Prior to plate tectonics, we made observations but could only guess at mechanisms. It was like watching the hands on a clock and trying to guess what moves them. After plate tectonics it was like being able to open the clock and not only watch the gears turn, but realize for the first time that there are such things as gears. Plate tectonics not only explains why things have happened, but also allows us to predict what might happen in the future.

Plate tectonics is covered in more detail later, however the key point is that Earth's outer layer consists of rigid plates that are constantly interacting with each other as they move around the Earth. The boundaries of plates move away from each other in some places, collide in others, and sometimes just slide past each other (illustrated by the red arrows in figure). The plates can move because they are floating on a layer of weak rock that deforms as the plates travel, much the same way the filling in a peanut butter and jelly sandwich allows you to slide the top layer of bread across the bottom layer.

Whether the plates move away from each other, collide, or just slide past each other determines things like the locations of mountain belts and volcanoes, where earthquakes happen, and the shapes and sizes of oceans and continents.

Geologic Materials

Geologic materials include both soils and rocks but a well-defined boundary between them does not exist. For example, a highly weathered rock shown as a rock on a geologic map may be considered soil by a civil engineer. A soil is defined as a loose, unconsolidated aggregation of mineral particles that can be easily separated by hand pressure or agitation in water and that can be excavated without blasting.

A rock, on the other hand, is a hard and compact aggregation of minerals that remains intact in water and cannot be excavated without blasting. Between these two end members, there is a gradation zone (glacial tills, shales, claystones, mudstones, etc.) that exhibits properties of both soils and rocks.

Soils

Geologically, soils are the end products of mechanical and/or chemical weathering of rocks. Based on their mode of origin, soils can be categorized as residual or transported. Residual soils are those that remain at the place of their origin whereas transported soils are the ones that have been carried away from their place of origin by such agents as gravity (colluvial soils), water (alluvial soils), ice (glacial soils), and wind (aeolian soils).

Engineering properties of soils are closely related to their mode of origin. The residual soils are likely to exhibit a well-developed soil profile, colluvial soils are dominated by angular particles, alluvial soils are generally stratified, glacial soils can be highly heterogeneous with wide range in particle size, and aeolian soils are characterized by uniform particle size.

Rocks

Rocks are classified as igneous, sedimentary, and metamorphic on the basis of their origin. Igneous rocks are formed by solidification of molten rock material (lava or magma) upon cooling, sedimentary rocks are deposited in water in the form of layers, and metamorphic rocks are formed by the action of heat and pressure on pre-existing rocks. Detailed descriptions and classifications of these three types of rock can be found in books on physical geology. The International Association of Engineering Geology (IAEG) has also developed comprehensive classifications for igneous, sedimentary, and metamorphic rocks which can be useful for predicting their engineering behavior.

Igneous Rocks

Igneous rocks originate from the cooling of molten rock material that is termed magma when it lies below the ground surface and lava when it extrudes on the surface. Cooling below the earth's surface is slow and results in large crystals whereas cooling on the surface is rapid and results in small crystal size. Depending upon the place of cooling, igneous rocks are divided into extrusive or volcanic rocks (cool on the surface; fine-grained), hypabyssal rocks (cool a short distance below the surface; medium-grained), and intrusive or plutonic rocks (cool deep down inside the earth; coarse-grained).

Igneous rocks are usually classified on the basis of their texture (size, shape, and arrangement of grains) and mineral composition. Commonly encountered igneous rocks include granite, diorite, and gabbro as the coarse-grained varieties, rhyolite, andesite, and basalt as the fine-grained equivalents, and dolerite or diabase as the mediumgrained equivalent of gabbro. Extrusive igneous rocks also include volcanic breccia and tuff that are made of pyroclastic material of varying sizes (material that is blown out in the air during volcanic eruptions).

Engineering properties of igneous rocks are related to their texture and mineral composition. Fine-grained acid igneous rocks containing volcanic glass, opal, and chalcedony can result in alkali-silica reaction, when used as aggregate in Portland cement concrete, leading to volumetric expansion and cracking.

Coarse-grained igneous rocks have generally lower strength and abrasion resistance than fine-grained igneous rocks and are, consequently, less suitable for engineering applications. The mode of emplacement of intrusive igneous rocks is also significant in terms of engineering characterization. Massive plutonic rocks (stocks and batholiths) tend to be more isotropic with respect to engineering properties and less problematic in engineering construction than tabular plutons (dikes and sills). Among the extrusive rocks, those formed of lava flows are expected to be more homogeneous and closely jointed than the pyroclastic rocks that are heterogeneous with widely spaced jointing.

Sedimentary Rocks

Sedimentary rocks comprise about 75% of the rocks exposed at the earth surface. These rocks are deposited as sediments in water that are later lithified through the processes of compaction, cementation, and crystallization to form the rock. Layering or stratification is the single most characteristic feature of sedimentary rocks. Based on their origin, sedimentary rocks are divided into clastic sedimentary rocks, comprised of material broken down from pre-existing rocks, and chemical sedimentary rocks resulting from chemical precipitation, both inorganic and biochemical. Breccias, conglomerates, sandstones, shales, clystones, mudstones, and siltstones are examples of clastic sedimentary rocks.

Limestones, dolomites, and evaporates belong to the family of chemical, crystalline, sedimentary rocks of inorganic nature while chalk and coal are good examples of organic sedimentary rocks. According to West, the three rock types that comprise 99% of all sedimentary rocks are shales (46%), sandstones (32%), and limestones (22%).

Sedimentary rocks are extremely diverse in their texture and mineral composition. The source rock, agent of transportation, duration and distance of transportation, depositional environment, lithification processes, and type and amount of cementation all contribute to the diversity of sedimentary rocks. The diverse nature of sedimentary rocks is manifested by the extreme variation in their engineering properties.

Alkali-carbonate reaction, low abrasion resistance of coarse limestone aggregate, susceptibility of argillaceous carbonates, shales, and cherts to pitting and popouts during freeze-thaw cycles, low strength and high swelling potential of some shales, claystones, and mudstones, and presence of solution cavities and open fractures in limestone, are some of the problems posed by sedimentary rocks in engineering works.

Metamorphic Rocks

Metamorphic rocks form by the action of heat and pressure on pre-existing rocks within the earth's crust. Metamorphism, which results in textural, structural, and mineralogical changes in the parent rock, can take place near igneous intrusions (contact metamorphism) or over very large areas associated with plate movements (regional metamorphism). Most regionally metamorphosed rocks exhibit foliated texture that results from arrangement of minerals in parallel planes under the influence of pressure. The foliated metamorphic rocks include slates, phyllites, schists, and gneisses. The nonfoliated metamorphic rocks include marble (metamorphosed limestone) and quartzite (metamorphosed sandstone).

Foliated metamorphic rocks exhibit directional properties. Strength, permeability, and seismic velocity are strongly affected by direction of foliation. Since foliation represents planes of weakness in metamorphic rocks, serious stability problems can arise in projects involving slopes, tunnels, and foundations.

Engineering Characterization of Soils

Soils constitute one of the most widely encountered materials in engineering construction. Many engineering structures are either made of soil material or they are founded on soils. The design and stability of these structures depends on the engineering properties of soils involved that, in turn, are greatly influenced by their geologic characteristics. Because of their heterogeneity, anisotropic nature, and non-linear stressstrain curves, characterization and prediction of the engineering behavior of soils is a challenging job that requires experience and good judgment. For characterization purposes, engineering properties of soils are commonly grouped into index properties and design properties.

Index Properties

Phase Relationships

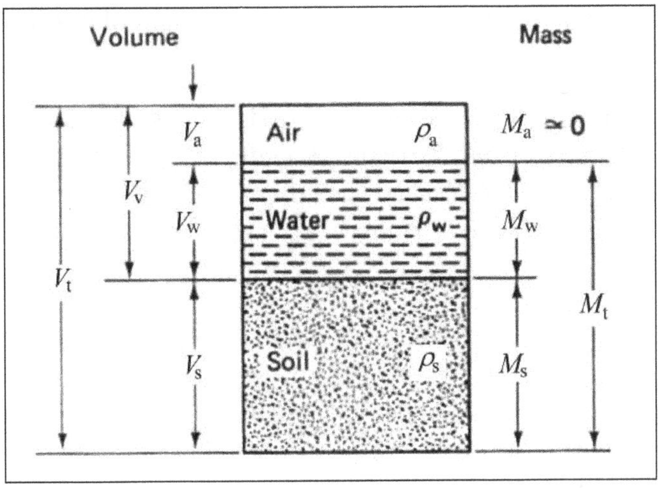

Phase diagram showing mass-volume relationships for soils.

1. A mass of soil commonly consists of three phases: solid mineral particles, water, and air. For a completely saturated and a completely dry soil, all voids (open spaces between solid particles) are filled with water and air respectively, and the soil mass reduces to a two-phase system. Figure shows a schematic representation of the masses and volumes of various phases involved. The inter-relationships between these phases define some important index properties used for soil characterization. The symbols used in defining various properties described below are given in figure.

2. Void Ratio (e): Void ratio is defined as the ratio of the volume of voids to the volume of solids, expressed as a decimal $(e = V_v / V_s)$. The higher the void ratio, the more compressible is the soil. Typical values of void ratio can range from 0.4-1.0 for sands, from 0.3-1.5 for clays, and much higher for organic soils.

3. Porosity (n): Porosity is the ratio of the volume of voids to the total volume of a soil mass, expressed as a percentage $\{n = (V_v / V_t)100\}$. Theoretically, porosity can range from 0-100%. Clayey soils tend to have higher porosity values (30-70%) than sandy soils (20-50%). Void ratio and porosity are related to each other as follows:

$$e = n/1-n$$

$$n = e/1+e$$

4. Degree of Saturation (S): Degree of saturation relates to the amount of water in the voids. It is the ratio of the volume of water to the volume of voids, expressed as a percentage $\{S = (V_w / V_v)100\}$. The degree of saturation ranges from 0% for a completely dry soil to 100% for a completely saturated soil. The lower the degree of saturation of an expansive clayey soil, the more will it expand upon the addition of water.

5. Water Content (w): Water content is the ratio of the mass of water to the mass of solids, expressed as a percentage $\{w = (M_w / M_s)100\}$. The water content for natural soils can range from 0% for a completely dry soil to several hundred percent for some marine organic clays. The higher the natural water content of a soil, the more undesirable are its engineering properties.

6. Density (ρ): Density connects the two sides of the phase diagram in Figure. Density is the ratio of the mass to the volume of a soil. Different types of density are used in engineering practice such as bulk density $(\rho = M_t / V_t)$, solid density $(\rho_s = M_s / V_s)$, dry density $(\rho_d = M_s / V_t)$, saturated density $\{\rho_{sat}(M_s + M_w) / V_t$, with M_w at $S = 100\%\}$, and submerged density $(\rho' = \rho_{sat} - \rho_w)$.

Soil Texture

Soil texture relates to particle sizes and shapes as well as distribution of various sizes (gradation) in a soil mass. Based on soil texture, soils are subdivided into coarse textured and fine-textured categories. Sands and gravels are considered coarse-textured soils and silts and clays fine-textured; the distinction between the two groups being whether the grains can be seen with the naked eye or not. Texture controls the behaviour of coarse-grained or granular soils and water controls the behaviour of fine-grained or cohesive soils.

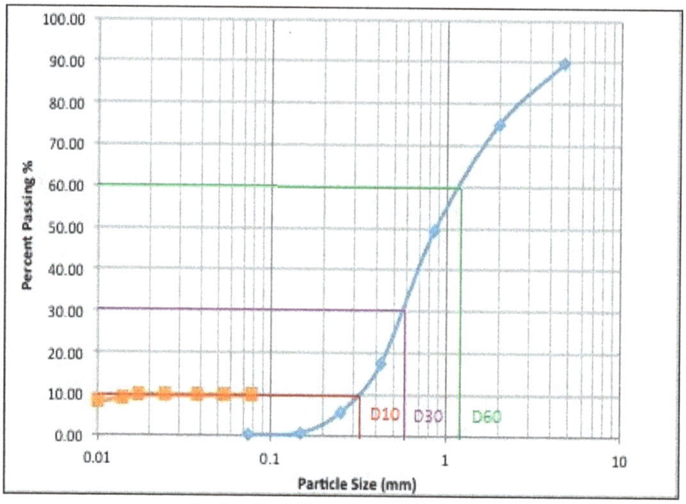

Grain size distribution curves.

Particle size can vary from boulders (10^3 mm) to colloidal size clay material (10^{-5}mm). Different grain size ranges are used for classification purposes by American Society for Testing and Materials (ASTM), American Association of State Highway and Transportation Officials (AASHTO), British Standards, and Unified Soil Classification System (USCS) details of which can be found in most soil mechanics books. Particle size distribution for coarse-grained soils is determined by performing a sieve analysis, using a set of standardized sieves.

The method has been standardized by ASTM and is designated as D 422. For the material passing the No. 200 sieve (0.074mm) and for fine-grained soils, the hydrometer analysis (ASTM D 422), based on Stokes' Law, is used. A grain size distribution curve is obtained by plotting the cumulative percent passing against the corresponding sieve sizes using a semi-log paper. Figure shows the grain size distribution curves for three different soils.

A well-graded soil is the one in which all particle sizes are well represented, a gapgraded soil has certain sizes missing, and a uniformly-graded or poorly-graded soil consists predominantly of one size particles. In engineering practice, a well-graded soil is expected to exhibit the best engineering properties whereas uniformly graded soils can be problematic.

Two quantitative indices, as defined below, are commonly used to describe the soil gradation:

$$\text{Coefficient of uniformity} = C_u = D_{60} / D_{10}$$

$$\text{Coefficient of curvature} = C_c = \left(D_{30}\right)^2 / \left(D_{10}\right)\left(D_{30}\right)$$

where D_{10}, D_{30}, and D_{60} are particle sizes corresponding to 10%, 30%, and 60% of the soil finer than the corresponding diameter respectively (on cumulative weight percent basis).

The smaller the C_u, the more uniformly graded the soil is. Well-graded soils have C_u values greater than 15. Alternatively, a soil will be well graded if its C_c is between 1 and 3, and C_u is greater than 4 for gravels and greater than 6 for sands.

Atterberg Limits

Atterberg limits, also referred to as consistency limits, represent water contents at which marked changes in physical state and engineering behavior of fine-grained soils occur. By comparing the natural water content of a soil with its Atterberg limits, one can predict its engineering behavior. Important Atterberg limits include liquid limit (LL), plastic limit (PL), and shrinkage limit (SL).

Liquid limit is the minimum water content at or above which a soil behaves as a viscous liquid and plastic limit is the minimum water content at which a soil behaves as a plastic material. Liquid and plastic limits for fine-grained soils can be determined by ASTM method D 4318. The numerical difference between the liquid and plastic limits is referred to as the plasticity index (PI).

It indicates the range of water content over which a soil behaves as a plastic material. Volume of a soil continues to decrease upon drying. Shrinkage limit is the minimum water content beyond which no further reduction in volume occurs upon further drying.

Although Atterberg limits are index properties, they are extremely important in characterization of fine-grained soils as they can be correlated with almost all other engineering properties. Soils with low SL and high PI values are most prone to detrimental volume change with changes in water content.

Composition and Structure of Earth

Core, mantle, and crust are divisions based on composition. The crust makes up less than 1 percent of Earth by mass, consisting of oceanic crust and continental crust is often more felsic rock. The mantle is hot and represents about 68 percent of Earth's mass. Finally, the core is mostly iron metal. The core makes up about 31% of the Earth. Lithosphere and asthenosphere are divisions based on mechanical properties. The lithosphere is composed of both the crust and the portion of the upper mantle that behaves as a brittle, rigid solid. The asthenosphere is partially molten upper mantle material that behaves plastically and can flow.

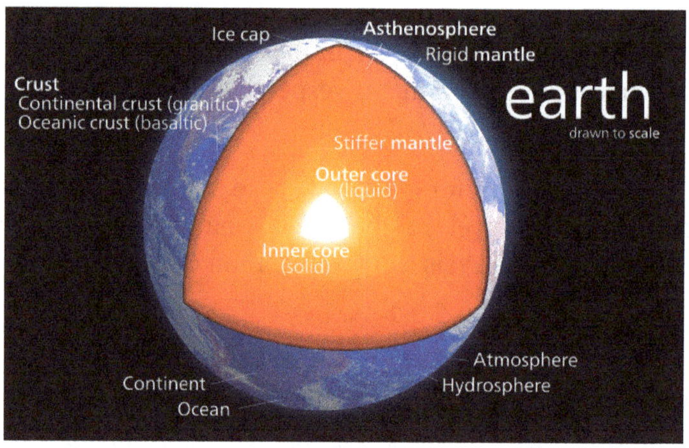

Crust and Lithosphere

Earth's outer surface is its crust; a cold, thin, brittle outer shell made of rock. The crust is very thin, relative to the radius of the planet. There are two very different types of crust, each with its own distinctive physical and chemical properties. Oceanic crust is composed of magma that erupts on the seafloor to create basalt lava flows or cools deeper down to create the intrusive igneous rock gabbro. Sediments, primarily muds and the shells of tiny sea creatures, coat the seafloor. Sediment is thickest near the shore where it comes off the continents in rivers and on wind currents. Continental crust is made up of many different types of igneous, metamorphic, and sedimentary rocks. The average composition is granite, which is much less dense than the mafic igneous rocks of the oceanic crust. Because it is thick and has relatively low density, continental crust rises higher on the mantle than oceanic crust, which sinks into the mantle to form basins.

When filled with water, these basins form the planet's oceans. The lithosphere is the outermost mechanical layer, which behaves as a brittle, rigid solid. The lithosphere is about 100 kilometers thick. The definition of the lithosphere is based on how earth materials behave, so it includes the crust and the uppermost mantle, which are both brittle. Since it is rigid and brittle, when stresses act on the lithosphere, it breaks. This is what we experience as an earthquake.

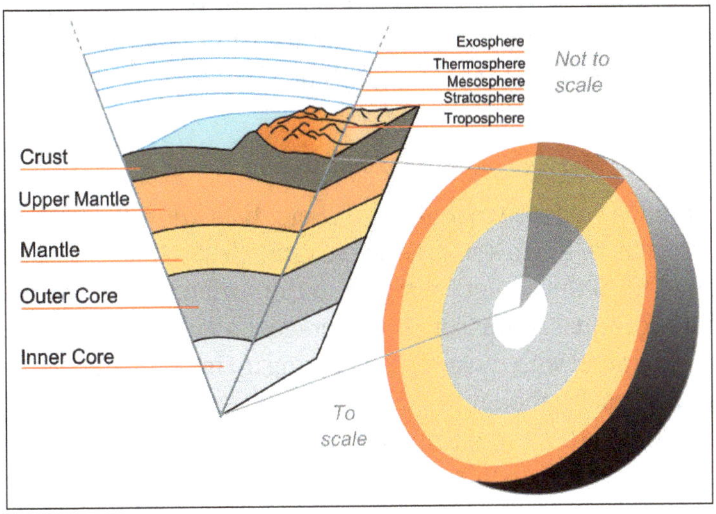

Mantle

The two most important things about the mantle are: (1) it is made of solid rock, and (2) it is hot. Scientists know that the mantle is made of rock based on evidence from seismic waves, heat flow, and meteorites. The properties fit the ultramafic rock peridotite, which is made of the iron- and magnesium-rich silicate minerals. Peridotite is rarely found at Earth's surface. Scientists know that the mantle is extremely hot because of the heat flowing outward from it and because of its physical properties. Heat flows in two different ways within the Earth: conduction and convection. Conduction is defined as the heat transfer that occurs through rapid collisions of atoms, which can only happen if the material is solid. Heat flows from warmer to cooler places until all are the same temperature. The mantle is hot mostly because of heat conducted from the core. Convection is the process of a material that can move and flow may develop convection currents. Convection in the mantle is the same as convection in a pot of water on a stove. Convection currents within Earth's mantle form as material near the core heats up. As the core heats the bottom layer of mantle material, particles move more rapidly, decreasing its density and causing it to rise. The rising material begins the convection current. When the warm material reaches the surface, it spreads horizontally. The material cools because it is no longer near the core. It eventually becomes cool and dense enough to sink back down into the mantle. At the bottom of the mantle, the material travels horizontally and is heated by the core. It reaches the location where warm mantle material rises, and the mantle convection cell is complete.

Convection in the mantle is the same as convection in a pot of water on a stove. Convection currents within Earth's mantle form as material near the core heats up. As the core heats the bottom layer of mantle material, particles move more rapidly, decreasing its density and causing it to rise. The rising material begins the convection current. When the warm material reaches the surface, it spreads horizontally. The material cools because it is no longer near the core. It eventually becomes cool and dense enough to sink back down into the mantle. At the bottom of the mantle, the material travels horizontally and is heated by the core. It reaches the location where warm mantle material rises, and the mantle convection cell is complete.

Core

At the planet's center lies a dense metallic core. Scientists know that the core is metal for a few reasons. The density of Earth's surface layers is much less than the overall density of the planet, as calculated from the planet's rotation. If the surface layers are less dense than average, then the interior must be denser than average. Calculations indicate that the core is about 85 percent iron metal with nickel metal making up much of the remaining 15 percent. Also, metallic meteorites are thought to be representative of the core. If Earth's core were not metal, the planet would not have a magnetic field. Metals such as iron are magnetic, but rock, which makes up the mantle and crust, is not. Scientists know that the outer core is liquid and the inner core is solid because S-waves stop at the inner core. The strong magnetic field is caused by convection in the liquid outer core. Convection currents in the outer core are due to heat from the even hotter inner core. The heat that keeps the outer core from solidifying is produced by the breakdown of radioactive elements in the inner core.

Geologic Time

The geological time scale shown in a simplified form is one of the crowning achievements of science in general and geology in particular. It is a reference and communication system for comparing rocks and fossils from throughout the world and is geology's equivalent of the periodic table of the elements.

Most of the boundaries on the geological time scale correspond to the origination or extinction of particular kinds of fossils. Knowing when major groups of fossils first appeared or went extinct is therefore incredibly useful for determining the ages of rocks in the field. For example, if you find a rock with a trilobite fossil upon it, you will immediately know that the rock is Paleozoic in age (541 Ma to 252 Ma) and not older or younger; knowing the species of trilobite allows even greater precision.

This relates to a third important principle of relative age dating: the principle of faunal succession. Faunal succession is the principle that different kinds of fossils characterize different intervals of time. This is because evolution and extinction are facts of nature.

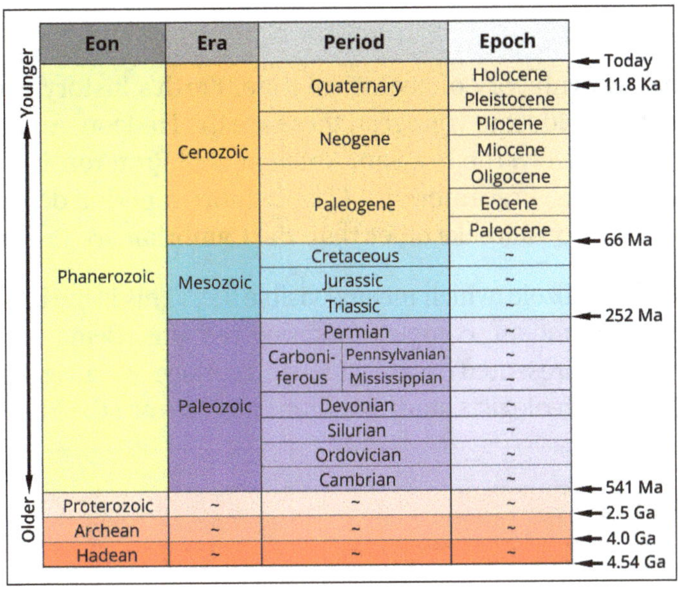

The geological time scale.

The principle of faunal succession was developed by an English surveyor named William "Strata" Smith. As he studied layers of rocks to determine where to build canals, he noticed that he found the same ordering of fossil species from place to place; Fossil A was always found below Fossil B, which in turn was always found below Fossil C, and so on. By documenting these sequences of fossils, Smith was able to temporally correlate rock layers (or, strata) from place to place (in other words, to establish that rock layers in two different places are equivalent in age based upon the fact that they include the same types of fossils). Temporal correlation allowed Smith to construct the first geological map of an entire country.

Temporal correlation has made many people very, very rich by allowing them to predict the locations of valuable geological resources such as fossil fuels. More generally, it has allowed us to reconstruct the geological history of Earth by comparing rocks and fossils from place to place. This is critically important because no single place on Earth preserves a complete geological history, or even a small fraction of it.

The geological time scale provides a global summary of countless small-scale temporal correlations of rock layers made at local and regional scales. It is based almost entirely upon careful observations of the distributions of fossils in time and space.

Learning the Geological Time Scale

Because of its usefulness for communicating about events in Earth's history, it is important that all students of geology, paleontology, and evolutionary biology commit the geological time scale to memory. This is most easily done by first breaking the time scale into its component parts: eons, eras, periods, and epochs.

Eons

The eon is the broadest category of geological time. Earth's history is characterized by four eons; in order from oldest to youngest, these are the Hadeon, Archean, Proterozoic, and Phanerozoic. Collectively, the Hadean, Archean, and Proterozoic are sometimes informally referred to as the "Precambrian." (The Cambrian period defines the beginning of the Phanerozoic eon; so, all rocks older than the Cambrian are Precambrian in age.)

We live during the Phanerozoic, which means "visible life." This is the interval of geological time characterized by abundant, complex fossilized remains. Being the youngest eon of time, it is also very well represented by rock at Earth's surface. Because of these two factors, most paleontologists and geologists study fossils and rocks from the Phanerozoic eon.

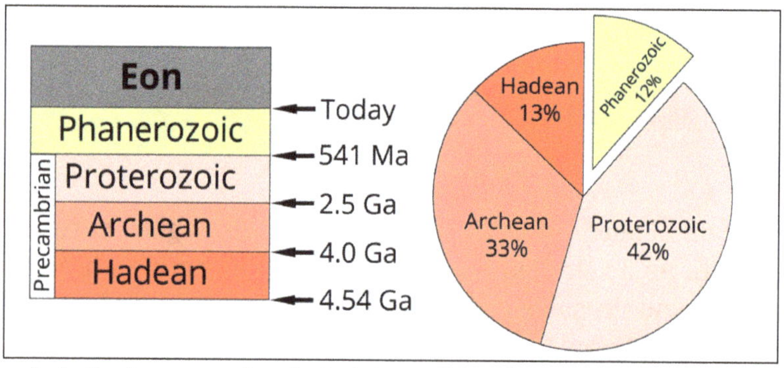

Left: the four eons of geological time. Right: the "Precambrian" eons (Hadean, Archean, and Proterozoic) represent 88% of geological time.

You have already learned that the Earth is 4.54 billion years old. The Phanerozoic eon began 541 million years ago (or, 0.541 billion years ago). Thus, the Phanerozoic eon represents a paltry 12% of Earth's history! Instead, most of Earth's history is represented by the three Precambrian eons. These older eons tell the story of Earth's beginning, life's origin, and the rise of complex life.

The Hadean and Archean are difficult eons to study, however, because they are exposed in very limited places on Earth's surface. (Since they are the oldest eons, rocks that are Hadean and Archean in age are often buried far below younger rocks at Earth's surface.) Proterozoic rocks which span nearly 2 billion years (42% of Earth's history) are much more accessible, but, until recently, have received significantly less attention from paleontologists than rocks from the younger, fossil-rich Phanerozoic eon. That is slowly beginning to change, however, as more clues about the origins of complex life begin to be revealed from Proterozoic-aged rocks.

Eras

Eons of geological time are subdivided into eras, which are the second-longest units of geological time. The Phanerozoic eon is divided into three eras: the Paleozoic, Mesozoic, and Cenozoic.

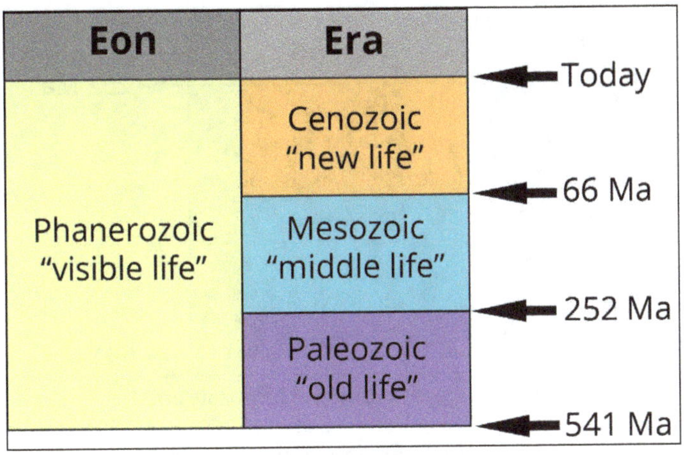

The three eras of the Phanerozoic eon.

Most of our knowledge of the fossil record comes from the three eras of the Phanerozoic eon. The Paleozoic ("old life") era is characterized by trilobites, the first four-limbed vertebrates, and the origin of land plants. The Mesozoic ("middle life") era represents the "age of dinosaurs," though also is noteworthy for the first appearances of mammals and flowering plants. Finally, the Cenozoic ("new life") era is sometimes called the "age of mammals" and is the era during which we live today.

As temporal points of reference, it is worth memorizing the ages of the boundaries that separate the three eras of the Phanerozoic eon. Long before geologists knew these absolute age dates, they realized that the boundaries represent important events in the history of life: mass extinctions. For example, many fossils that are commonly found in the youngest Paleozoic rocks are not found in overlying Mesozoic rocks. Similarly, dinosaur fossils found in the youngest Mesozoic rocks are never again found in the overlying Cenozoic rocks. Paleontologists and geologists used these mass extinction events to define these (and other) boundaries within the Phanerozoic portion of the geological time scale. It is therefore no coincidence that some of the major boundaries coincide with mass extinction events.

The older Archean and Proterozoic eons are similarly divided into several eras. For example, the youngest era of the Proterozoic eon is called the Neoproterozoic.

Periods

Just as eons are subdivided into eras, eras are subdivided into units of time called periods. The most well known of all geological periods is the Jurassic period of the Mesozoic era.

The Paleozoic era is divided into six periods. From oldest to youngest, these are the Cambrian, Ordovician, Silurian, Devonian, Carboniferous, and Permian. Note that in the United States, the Carboniferous is divided into two separate periods: the Mississippian and the Pennsylvanian.

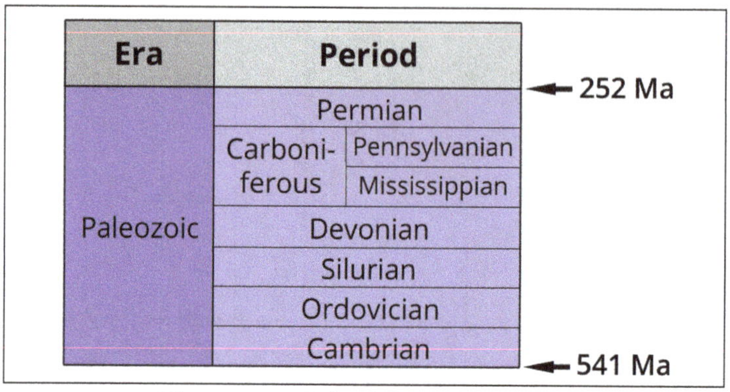

The seven periods of the Paleozoic era.

The Mesozoic era is divided into the Triassic, Jurassic, and Cretaceous periods.

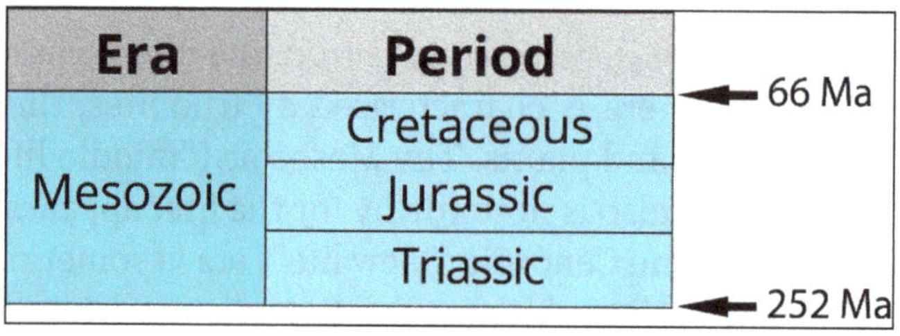

The three periods of the Mesozoic era.

Finally, the Cenozoic era is divided into three periods: the Paleogene, Neogene, and Quaternary.

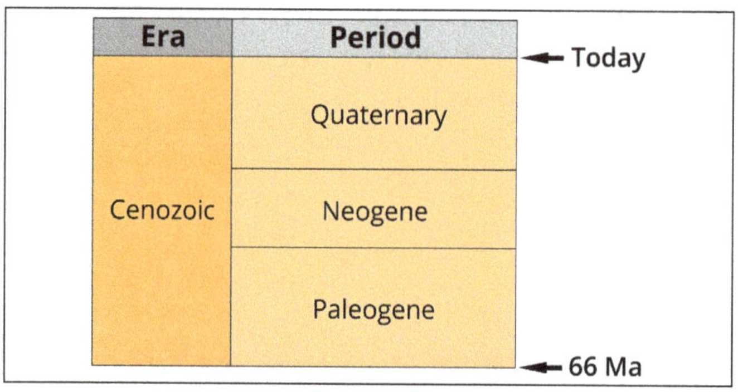

The three periods of the Cenozoic era.

Epochs and Ages

Periods of geological time are subdivided into epochs. In turn, epochs are divided into even narrower units of time called ages. It is important to note, however, that all of the periods of the Phanerozoic era are subdivided into the epochs and ages.

The Paleogene period is divided into from oldest to youngest the Paleocene, Eocene, and Oligocene epochs. The Neogene is divided into the Miocene and Pliocene epochs. Finally, the Quaternary is divided into the Pleistocene and Holocene epochs. Some geologists now think that since humans are having such a notable impact on the Earth and its life a new, youngest epoch should be added to the Quaternary the Anthropocene. There is still considerable discussion in the geological community about whether this epoch should be added, as well as debate about what characteristics should define its beginning.

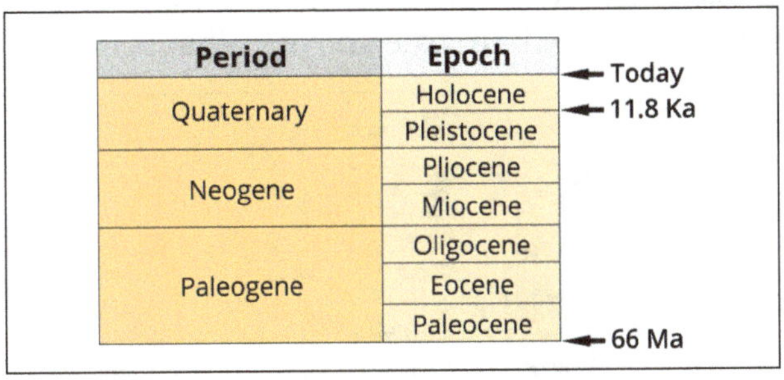

Epochs of the Paleogene, Neogene and Quaternary periods.

The Geological Time Scale

Now that you have learned about the hierarchical components of the geological time scale eons, eras, periods, and epochs consider again how all of these parts fit together. Note that some boundaries (those that follow horizontal lines on the time scale) are equivalent in age. For example, the Mesozoic-Cenozoic boundary is equivalent to the Cretaceous-Paleogene boundary (both have an age of 66 Ma). Similarly, the Paleogene-Neogene boundary is equivalent to the Oligocene-Miocene boundary.

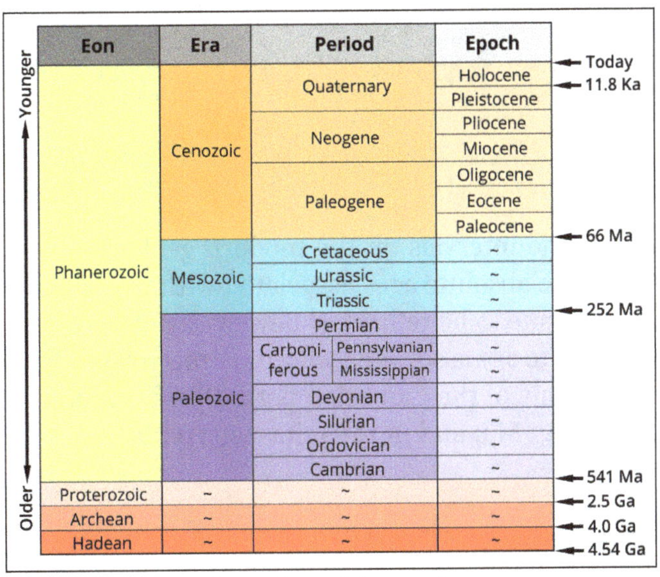

The geological time scale.

It is much easier to memorize the time scale by first breaking it down into its component parts. Many geology students have developed clever mnemonic devices to help remember all of the names in the time scale and a quick Google search will provide many examples (which are often rather salty!) that you may find helpful as you memorize the time scale.

The geological time scale is a result of hundreds of years of investigation and remains very much a work in progress. Observe the geological time scale below, which is from a college-level geology textbook published over 130 years ago by Le Conte (1885). What similarities does it share with the time scale above? What is different?

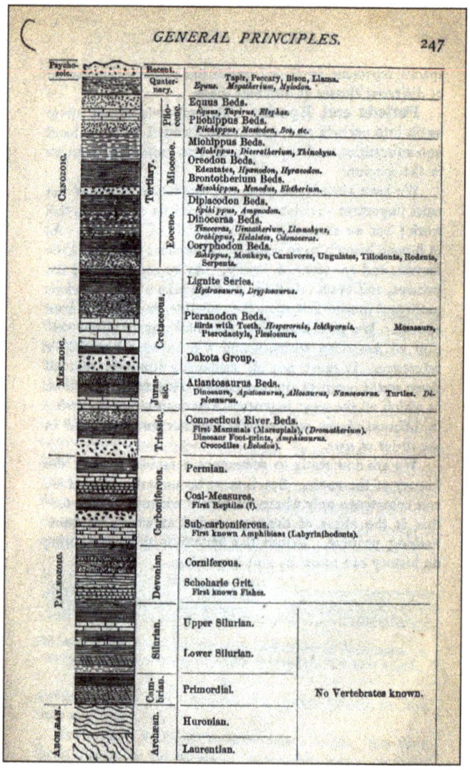

Geological time scale from Le Conte (1885).

Most of the names of time intervals on the modern geological time scale have been stable for many years. The absolute ages associated with the boundaries that separate these time intervals continue to be refined as new evidence becomes available. For example, the modern absolute age associated with the Proterozoic-Phanerozoic boundary (equivalent to the beginning of the Cambrian period) is 541 Ma. In 2009, this age was 542 Ma; in 1999, it was 543 Ma; and in 1983, it was 570 Ma.

Geologic Dating

For centuries people have argued about the age of the Earth; only recently has it been possible to come close to achieving reliable estimates. In the 19th century some geologists realized that the vast thicknesses of sedimentary rocks meant that the Earth must be at least hundreds of millions of years old. Charles Darwin reinforced this idea by pointing to the time that must have been required for the evolution of advanced life from primitive forms.

On the other hand, the great physicist Lord Kelvin vehemently objected and suggested that the Earth might only be a few tens of millions of years old, based on his calculations of its cooling history. These discussions were rendered obsolete by the discovery of radioactivity in 1896 by the French physicist Henri Becquerel. The existence of radio activities of various kinds in rocks has enabled earth scientists to determine the age of the Earth, the moon, meteorites, mountain chains and ocean basins, and to draw up a reasonably accurate time scale of evolution. It has even been possible to work out a time scale of the reversals of the Earth's magnetic field. This "radiometric" approach has superseded all other techniques for determining absolute ages.

Radioactive Clock

The vast majority of atoms (each composed of a nucleus surrounded by electrons) are stable. Essentially, they will exist forever. A critical few, however, are unstable. Their nuclei tend to emit particles spontaneously - i.e., they are radioactive. Because of this particle emission, the original radioactive parent atom changes its identity, becoming a different, stable daughter atom. This change takes place at a known rate determined by the half-life; ie, the time required for one-half of the original number of radioactive atoms to convert to the stable daughter product. The remaining number of radioactive atoms is halved every half-life. Radioactive elements of use in geological dating have relatively long half-lives. A good example is rubidium-87, which changes to strontium-87 at a rate of one-half every 50 billion years. Therefore, a rock can be dated by measuring how much of its original rubidium content has changed into strontium.

The other key dating techniques involve uranium-235 transforming to lead-207 at a rate of one-half every 713 million years, uranium-238 becoming lead-206 at one-half every 4.5 billion years, potassium changing to argon (and calcium) at one-half every 1.3 billion years and samarium-147 becoming neodymium-143 at one-half every 106 billion years. These radioactive processes present a set of natural clocks which reveal when the rock was formed, or when it was last heated severely. The well-known carbon-14 method involves the conversion of radioactive carbon-14 to stable nitrogen at a rate of one-half about every 5700 years. It can only be used to date organic matter, and is accurate only for materials younger than about 50 000 years.

Since 1950, radiometric methods have been developed to a very sophisticated level in several countries, including Canada. It has been demonstrated that when rocks which have led an undisturbed history are analysed, all methods reveal the same age. This uniformity demonstrates that the principle is reliable. When disturbed rocks are studied, the different techniques may give different readings, and much research has been carried out on how to interpret such results. It often proves possible to date even severely disturbed rocks.

Age of the Earth

To date the time of formation of a planet 12 740 km in diameter and 70.8% covered by water is not easy. Only the tiniest fraction of the Earth, the crust, is accessible. Those rocks available for analysis (ie, the oldest ones) have been heated and squeezed many times in their geological history, because for billions of years continents have been drifting over the Earth's surface, colliding and producing mountains and new ocean floors.

Two approaches have been developed to circumvent these problems. The first involves sampling as much of the Earth's crust as possible and dating these rocks. The Earth certainly must be older than the oldest terrestrial rocks found. A small area of metamorphic rock in northern Canada, known as the Acasta gneiss, is the oldest known intact solid piece of the Earth's crust. Using the uranium-lead technique they dated zircon crystals from the gneiss (located southeast of Great Bear Lake in the NWT) and showed that it was formed almost 4 billion years ago. Therefore it is clear that the Earth is over 4 billion years old. Consistent with this are the results of Stephen Moorbath and his colleagues at Oxford, who have shown that rocks near Godthaab in southwest Greenland either formed or were in existence approximately 3.8 billion years ago. These results have been confirmed and agreement has been found among the rubidium-strontium, uranium-lead and samarium-neodymium methods.

Rocks of almost this age have also been identified in other localities, including Labrador, Minnesota, Africa and India. Many scientists are searching for rocks older than these, and in 1983 Australian scientists claimed to have discovered minute zircon crystals 4.2 billion years old. They were found, however, in much younger sediment and it is not known where these zircons originated.

The second approach, which is more indirect but gives an answer currently believed correct, involves a comparison of the Earth with meteorites. They have clearly fallen to Earth from outside, often gouging out huge craters such as that called New Québec (61°17′ N, 73° 41' W). Rubidium-strontium, potassium-argon, uranium-lead and samarium-neodymium dating all show that the meteorites formed about 4.6 billion years ago. But detailed studies of lead isotopes in meteorites and terrestrial rocks strongly indicate that the Earth and meteorites formed at the same time.

Therefore, since the meteorites are very accurately dated at 4.6 billion years old, the Earth is also considered to be the same age. Dating of the lunar samples collected by the Apollo missions strongly indicates that the moon is of the same age. If the Earth, the moon and meteorites are all 4.6 billion years old, then so very probably is the whole solar system.

Age of Canadian Shield

The most ancient rocks of Canada comprise the Canadian shield. Many Canadian scientists, and others, have examined these rocks to work out the region's history. The Shield is made up of areas of rocks of distinctive ages. Apart from the Acasta gneiss, the oldest are found in Saglek Bay and near Hebron Fjord in Labrador and are about 3.6 billion years old. Other massive slabs are dated at 2.9-2.5 billion years, 1.8-1.7 billion years and 1.3-0.9 billion years.

Some of these areas represent the roots of what were ancient mountain chains, the upper parts of which were long ago removed by erosion. Others represent volcanic belts, many of which have never been very deeply buried. Scientists are studying whether these portions of different age and geological history were always close together or were far apart until gathered together by continental drifting and plate tectonics. Although the igneous rocks of the Shield are very ancient, the formation of igneous rocks has been a continuing process in Canada. The rocks formed by the Aiyansh lava flow in BC are thought to be 90 to 350 years old, confirming legends of the Tsimshian people of the Nass River describing volcanic activity.

Time Scale of Biological Evolution

It seems probable that life has existed on the Earth for well over 3 billion years. FOSSIL bacteria have been tentatively identified in the Fig-Tree Sediments in South Africa. Volcanic rocks associated with these have been dated at 3.5 billion years by the samarium-neodymium method by scientists at Columbia University and by the potassium-argon method at the University of Toronto. Probably the oldest fossils in Canada (2.5 billion years old) are the stromatolite formations at Steep Rock Lake, Ont. An indisputably biogenic, highly diverse microfossil assemblage is present in the approximately 1.9-billion-year-old Gunflint cherts of southern Ontario. However, the fossil record is well documented only for the last 545 million years, for only during that period did organisms exist with the hard phosphate or calcium-carbonate components which make for good fossil preservation.

The final 545 million years of evolution have been divided by palaeontologists into 3 eras: Palaeozoic (ancient life), Mesozoic (middle life) and Cenozoic (present life). The eras are subdivided into periods. The numerical estimates of the time occupied by these periods were made mostly with the aid of potassium-argon, rubidium-strontium and uranium-lead dating of rocks, which could be correlated with the time scale.

Humans are creatures of only the last few million years. Fragmentary fossils of apelike, upright-walking potential human ancestors (Australopithecus anamensis) were found in 1995 in northern Kenya by an international expedition led by Meave Leakey of the National Museums of Kenya. Volcanic ash associated with these fossils has been dated at the Australian National University using a laser-probe dating technique developed at U of T. It is clear that creatures even remotely resembling modern humans have been on the Earth for less than one-tenth of 1% of our planet's history.

Dating Rocks and Fossils using Geologic Methods

Despite seeming like a relatively stable place, the Earth's surface has changed dramatically over the past 4.6 billion years. Mountains have been built and eroded, continents and oceans have moved great distances, and the Earth has fluctuated from being extremely cold and almost completely covered with ice to being very warm and ice-free. These changes typically occur so slowly that they are barely detectable over the span of a human life, yet even at this instant, the Earth's surface is moving and changing. As these changes have occurred, organisms have evolved, and remnants of some have been preserved as fossils.

A fossil can be studied to determine what kind of organism it represents, how the organism lived, and how it was preserved. However, by itself a fossil has little meaning unless it is placed within some context. The age of the fossil must be determined so it can be compared to other fossil species from the same time period. Understanding the ages of related fossil species helps scientists piece together the evolutionary history of a group of organisms.

For example, based on the primate fossil record, scientists know that living primates evolved from fossil primates and that this evolutionary history took tens of millions of years. By comparing fossils of different primate species, scientists can examine how features changed and how primates evolved through time. However, the age of each fossil primate needs to be determined so that fossils of the same age found in different parts of the world and fossils of different ages can be compared.

There are three general approaches that allow scientists to date geological materials and answer the question: "How old is this fossil?" First, the relative age of a fossil can be determined. Relative dating puts geologic events in chronological order without requiring that a specific numerical age be assigned to each event. Second, it is possible to determine the numerical age for fossils or earth materials. Numerical ages estimate the date of a geological event and can sometimes reveal quite precisely when a fossil species existed in time. Third, magnetism in rocks can be used to estimate the age of a fossil site. This method uses the orientation of the Earth's magnetic field, which has changed through time, to determine ages for fossils and rocks.

Relative Dating to Determine the Age of Rocks and Fossils

Geologists have established a set of principles that can be applied to sedimentary and volcanic rocks that are exposed at the Earth's surface to determine the relative ages of geological events preserved in the rock record. For example, in the rocks exposed in the walls of the Grand Canyon there are many horizontal layers, which are called strata. The study of strata is called stratigraphy, and using a few basic principles, it is possible to work out the relative ages of rocks.

Individual rock layers, or strata, can be seen exposed in the wall of the Grand Canyon in Arizona, USA.

Just as when they were deposited, the strata are mostly horizontal (principle of original horizontality). The layers of rock at the base of the canyon were deposited first, and are thus older than the layers of rock exposed at the top (principle of superposition).

In the Grand Canyon, the layers of strata are nearly horizontal. Most sediment is either laid down horizontally in bodies of water like the oceans, or on land on the margins of streams and rivers. Each time a new layer of sediment is deposited it is laid down horizontally on top of an older layer. This is the principle of original horizontality: layers of strata are deposited horizontally or nearly horizontally. Thus, any deformations of strata must have occurred after the rock was deposited.

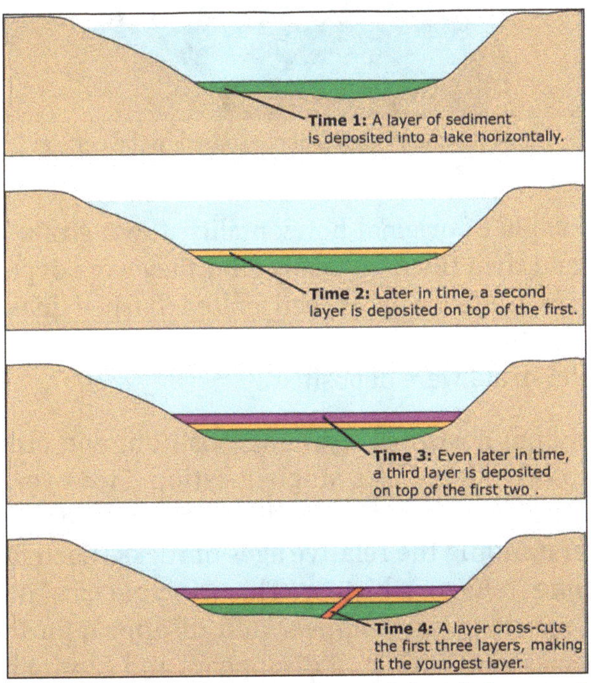

The principles of stratigraphy help us understand the relative age of rock layers.

Layers of rock are deposited horizontally at the bottom of a lake (principle of original horizontality). Younger layers are deposited on top of older layers (principle of superposition). Layers that cut across other layers are younger than the layers they cut through (principle of cross-cutting relationships).

The principle of superposition builds on the principle of original horizontality. The principle of superposition states that in an undeformed sequence of sedimentary rocks, each layer of rock is older than the one above it and younger than the one below it. Accordingly, the oldest rocks in a sequence are at the bottom and the youngest rocks are at the top.

Sometimes sedimentary rocks are disturbed by events, such as fault movements, that cut across layers after the rocks were deposited. This is the principle of cross-cutting relationships. The principle states that any geologic features that cut across strata must have formed after the rocks they cut through.

The sedimentary rock layers exposed in the cliffs at Zumaia, Spain, are now tilted close to vertical.

According to the principle of original horizontality, these strata must have been deposited horizontally and then titled vertically after they were deposited. In addition to being tilted horizontally, the layers have been faulted (dashed lines on figure). Applying the principle of cross-cutting relationships, this fault that offsets the layers of rock must have occurred after the strata were deposited.

The principles of original horizontality, superposition, and cross-cutting relationships allow events to be ordered at a single location. However, they do not reveal the relative ages of rocks preserved in two different areas. In this case, fossils can be useful tools for understanding the relative ages of rocks. Each fossil species reflects a unique period of time in Earth's history. The principle of faunal succession states that different fossil species always appear and disappear in the same order, and that once a fossil species goes extinct, it disappears and cannot reappear in younger rocks.

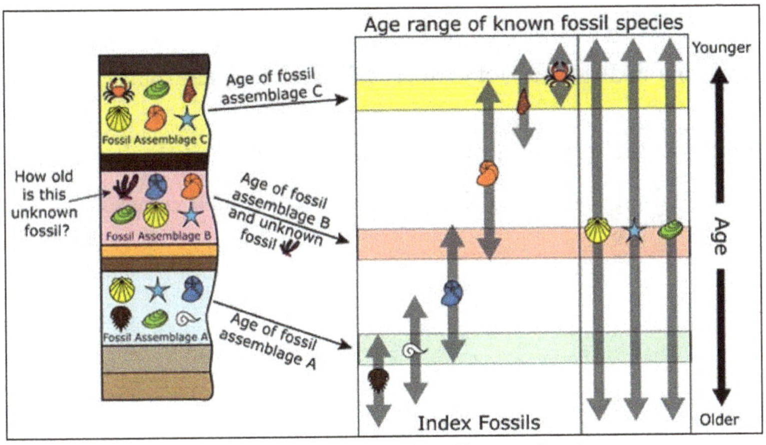

The principle of faunal succession allows scientists to use the fossils to understand the relative age of rocks and fossils.

Fossils occur for a distinct, limited interval of time. In the figure, that distinct age range for each fossil species is indicated by the grey arrows underlying the picture of each fossil. The position of the lower arrowhead indicates the first occurrence of the fossil and the upper arrowhead indicates its last occurrence – when it went extinct. Using the overlapping age ranges of multiple fossils, it is possible to determine the relative age of the fossil species (i.e., the relative interval of time during which that fossil species occurred). For example, there is a specific interval of time, indicated by the red box, during which both the blue ammonite and orange ammonite co-existed. If both the blue and orange ammonites are found together, the rock must have been deposited during the time interval indicated by the red box, which represents the time during which both fossil species co-existed. In this figure, the unknown fossil, a red sponge, occurs with five other fossils in fossil assemblage B. Fossil assemblage B includes the index fossils the orange ammonite and the blue ammonite, meaning that assemblage B must have been deposited during the interval of time indicated by the red box. Because, the unknown fossil, the red sponge, was found with the fossils in fossil assemblage B it also must have existed during the interval of time indicated by the red box.

Fossil species that are used to distinguish one layer from another are called index fossils. Index fossils occur for a limited interval of time. Usually index fossils are fossil organisms that are common, easily identified, and found across a large area. Because they are often rare, primate fossils are not usually good index fossils. Organisms like pigs and rodents are more typically used because they are more common, widely distributed, and evolve relatively rapidly.

Using the principle of faunal succession, if an unidentified fossil is found in the same rock layer as an index fossil, the two species must have existed during the same period of time. If the same index fossil is found in different areas, the strata in each area were likely deposited at the same time. Thus, the principle of faunal succession makes it possible to determine the relative age of unknown fossils and correlate fossil sites across large discontinuous areas.

References

- The-composition-and-structure-of-earth, geophysical: lumenlearning.com, Retrieved 05, August 2020
- Geological-time-scale, geological-time: digitalatlasofancientlife.org, Retrieved 11, January 2020
- Dating-rocks-and-fossils-using-geologic-methods-107924044: nature.com, Retrieved 20, May 2020

Chapter 2

Petrology and its Types

Petrology focuses on the rocks and the conditions under which they form. Petrology has three major types: igneous, metamorphic and sedimentary petrology. Igneous petrology is the study of physical and chemical properties of igneous rocks while metamorphic petrology is the study of composition and texture of metamorphic rocks. All the diverse principles of petrology and its types have been carefully analyzed in this chapter.

Petrology refers to the scientific study of rocks and the conditions which influence their formation. Petrology is a branch of geology that focuses on the chemical analysis in various fields such as petrography and mineralogy. By incorporating various principles of geophysics and geochemistry, modern petrologists can establish the origins of rocks and their chemical characteristics. There are three main branches of petrology resultant from the main rock types: sedimentary, igneous, and metamorphic.

Importance of Petrology

Petrology plays an important role in ascertaining the physical and chemical composition of rocks and the different conditions that influence their formation. Modern petrologists rely on knowledge in mineralogy to help in mapping and sampling of rocks. Since most rocks constitute various minerals, it becomes easier to study and understand them with background knowledge in mineralogy.

Petrological research is also crucial in helping us understand the nature of the earth's crust and mantle. Over many years, the earth's tectonic processes have shaped the nature of rocks and the topography of the earth's surface. These geological processes are vital in determining the suitability of certain areas for agriculture, industrial or commercial use.

Aside from helping us understand the history of our planet's formation, petrology applies in many areas that directly affect our livelihood. The field of petrology helps us to understand the best raw materials to be used in industries for the manufacture of goods. Additionally, petrology promotes best practices that foster sustenance and technological advancement.

Branches of Petrology

There are three main branches of petrology, namely: igneous, sedimentary, and metamorphic.

Igneous Petrology

Igneous petrology is a branch that specializes in the scientific study of igneous rocks, their chemical composition and texture. Igneous rocks are rocks formed when magma or molten rock is crystallized to form granite or basalt. Igneous rocks may be formed through crystallization or may involve a different process that cools or solidifies the molten rock. Igneous rocks that are formed beneath the earth's surface form 'intrusive' rocks, also known as plutonic rocks. Rocks found on the earth's surface produce extrusive rocks. About 95 percent of the earth's crust is made up of igneous rocks. However, these rocks do not readily appear on the surface because they are covered by a thin extensive layer of sedimentary and metamorphic rocks.

Sedimentary Petrology

This category focuses on the study of sedimentary rocks with an emphasis on texture and composition. Sedimentary rocks include shale, limestone, and sandstone. They are formed when particles derived from other rocks are bound together to form a matrix. These pieces of rock particles together with deposits of chemical and biological properties coalesce to produce a fine rock material.

Metamorphic Petrology

Metamorphic petrology involves the study of the composition and texture of metamorphic rocks. Such rocks include marble, gneiss, slate, and schist. Initially, sedimentary or igneous rocks undergo mineralogical, textural and chemical changes because of extreme pressure or temperature, and in some cases due to both. Due to high pressure and temperatures, the igneous or sedimentary rocks are transformed into metamorphic rocks.

Experimental Petrology

Experimental petrology involves the study of natural or synthetic materials when subjected to conditions of high pressure and high temperature. This helps to establish the phase relations and geochemical properties of the materials. The work of experimental petrology has provided a firm foundation for understanding the processes involved in the formation of metamorphic rocks. Rocks on the lower crust and upper mantle of the earth can be easily studied and analyzed as a result of experimental petrology.

The Practical Application of Petrology

Petrology and its application are important for several reasons:

- The study of rocks helps us to understand the composition of the earth's crust. The various minerals and chemical properties of the mantle are important to petrologist. Various professions such as civil engineering rely heavily on the findings of petrologists.

- By studying rocks, historical formation of the earth's surface can be established. Methods such as radiometric dating can be used to determine the time periods in which certain rock types were formed. This helps in creating a comprehensive series of geological events.
- Scientists can now understand the tectonic processes that led to the formation of the earth's surface. This is because specific tectonic environments have distinct characteristics from others. Scientists can, therefore, reconstitute these processes.
- Rocks are natural sources for important minerals and ores which are used as raw materials by industries. Petrologists help in the discovery of new minerals hidden below the earth's surface. The supply of such minerals is also vital to ensuring sustenance and commercial viability.
- The study of the earth's rocks helps in technological advancement.

Igneous Petrology

Igneous petrology is concerned with the identification, classification, origin, evolution, and processes of formation and crystallization of the igneous rocks.

Continental Igneous Rocks

A wide variety of igneous rocks occur in the continental lithosphere, a reflection of its heterogeneous nature compared to oceanic lithosphere. Because the continents are not subducted and are subject to uplift and erosion, older plutonic rocks are both preserved and accessible to study.

Granitic Rocks

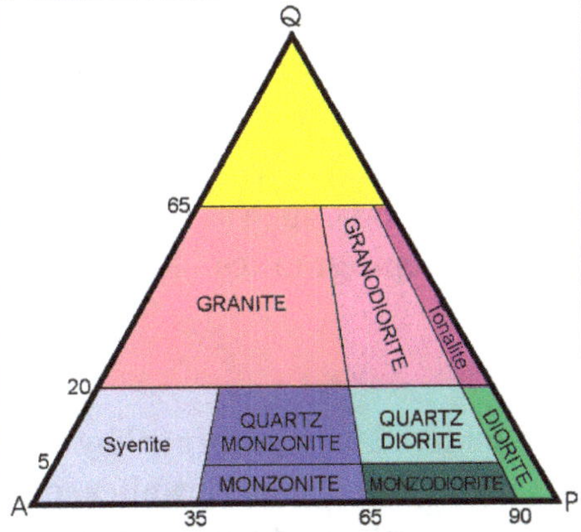

Highly Modified IUGS Classification of Phaneritic Igneous Rocks

- Classification of Granitic rocks based on visible pewrcentages of "Q" quartz / "A" alkalai feldspar/"P" plagioclase feldspar.

True granites have between 10% and 65% of their feldspars as plagioclase, and between 20% and 60% quartz. All rocks will likely contain mafic minerals such as biotite, hornblende, and perhaps pyroxenes, along with opaque oxide minerals.

"S", "I" and "A" Type Granitic Rocks

Mechanisms by which large volumes of granitic magma could be produced are:

- Anatexis of metasedimentary/sedimentary rocks to form S-type granitic magmas.
- Anatexis of young crustal basic meta-igneous rocks to form I-type granitic magmas.
- Melting/Assimilation of lower crustal rocks by mantle-derived basic magmas.
- Crystal fractionation/Assimilation of basaltic and andesitic magmas.
- Granitization, wherein high grade metamorphism bordering on melting converts rocks into those that appear texturally and mineralogically similar to granitic rocks.

S-type Granites

S-type granites are thought to originate by melting (or perhaps by ultrametamorphism) of a pre-exiting metasedimentary or sedimentary source rock. These are peraluminous granites [i.e. they have molecular Al_2O_3 ($Na_2O + K_2O$)].

Mineralogically this chemical condition is expressed by the presence of a peraluminous mineral, commonly muscovite, although other minerals such as the Al_2SiO_5 minerals and corundum may also occur. Since many sedimentary rocks are enriched in Al_2O_3 as a result of their constituents having been exposed to chemical weathering near the Earth's surface (particularly rocks such as shales that contain clay minerals), melting of these rocks is a simple way of achieving the peraluminous condition. Many S-type granitoids are found in the deeply eroded cores of fold-thrust mountain belts formed as a result of continent-continent collisions, such as the Himalayas and the Appalachians, and would thus be considered orogenic granites.

I-type Granites

I-type granites are granites considered to have formed by melting of an original igneous type source. These are generally metaluminous granites, expressed mineralogically by the absence of peraluminous minerals, and the absence of peralkaline minerals, Instead these rocks contain biotite and hornblende as the major mafic minerals.

Mesozoic or younger examples of I-type granites are found along continental margins such as the Sierra Nevada batholith of California and Nevada, and the Idaho batholith of Montana. In these regions the plutonism may have been related to active subduction beneath the western U.S. in Mesozoic times. I-type granites are also found in the Himalayas, related to continent-continent collisions.

Plutonic suites that were emplaced in convergent continental margin settings show many of the same characteristics as the calc-alkaline volcanic suite that likely erupted on the surface above. The suites include gabbros, diorites, quartz monzonites, granodiorites, and granites. They show only mild to no Fe-enrichment, similar to calc-alkaline volcanic rocks, and a range of isotopic compositions similar to the associated volcanic rocks. Nearly all are I-type granitoids.

The rocks were emplaced during the Mesozoic Era. Exposed rocks are generally older toward the east and southeast. Kistler and Peterman showed that the Sr isotopic ratios vary across the batholith in a systematic way. The younger rocks in the western portion of the batholith are mostly quartz diorites with Sr isotopic ratios less than 0.704, ratios expected from melting of the mantle or young crustal rocks. Plutons farther east are mostly quartz monzonites and granodiorites with ratios increasing along with age of the plutons toward the east and southeast. One interpretation of the data is that the older rocks contain a higher proportion of older crustal material than the younger plutonic bodies.

A-type Granites

A-type granites are generally peralkaline in composition [molecular $(Na_2O+K_2O)>Al_2O_3$]. Minerals like the sodic amphiboles riebeckite and arfvedsonite and the sodic pyroxene aegerine are commonly found in these rocks. In addition, they tend to be relatively Fe-rich and thus fayalitic olivine may sometimes occur.

These are considered anorogenic granites because they are generally found in areas that have not undergone mountain building events.Instead, they appear to be related to continental rifting events wherein continental lithosphere is thinned as a result of upwelling asthenosphere.The upwelling raises the geothermal gradient resulting in melting. Young peralkaline granites are found in the Basin and Range Province of the western U.S., and older examples are found throughout southeastern Australia.

Depth of Emplacement

Because the conditions under which a magma cools can play an important role in the texture and contact relationships observed in the final rock, plutons can be characterized by the depth at which they were emplaced. This is because depth, to a large extent controls the contrast in temperature between the magma and its surroundings.

Catazonal Plutons

The catazone is the deepest level of emplacement, usually considered to be at depths greater than about 11 km. In such an environment there is a low contrast in temperature between the magma and the surrounding country rock. The country rock itself is generally high grade metamorphic rock. Contacts between the plutons and the country rock are concordant (meaning the contacts run parallel to structures such as foliation in the surrounding country rock) and often gradational. The plutons themselves often show a foliation that is concordant with that in the surrounding metamorphic rocks. Migmatites (small pods of what appears to have been melted rock surrounded by and grading into metamorphic rocks) are common. Some catazonal plutons appear to have formed by either melting in place or by ultrametamorphism that grades into actually melting. Others appear to have intruded into ductile crustal rocks. Most, but not all, Catazonal plutons are S-type granitoids.

Mesozonal Plutons

The metazone occurs at intermediate crustal depths, likely between 8 and 12 km. The plutonic rocks are more easily distinguished from the surrounding metamorphic rocks. Contacts are both sharp and discordant (cutting across structures in the country rock), and gradational and concordant like in the catazone. Angular blocks of the surrounding country rock commonly occur within the plutons near their contacts with the country rock. The plutons generally lack foliation and are often chemically and mineralogically zoned.

Epizonal Plutons

The epizone is the shallowest zone of emplacement, probably within a few kilometers of the surface. In such an environment there is a large contrast between the temperatures of the magma and the country rock. The country rock is commonly metamorphosed, but the metamorphism is contact metamorphism produced by the heat of the intrusion. Contacts between the plutons and surrounding country rock are sharp and discordant, indicating intrusion into brittle and cooler crust. The margins of the plutons often contain abundant xenoliths of the country rock.

Pegmatities

Pegmatites are very coarse grained felsic rocks that occur as dikes or pod-like segregations both within granitic plutons and intruded into the surrounding country rock. They appear to form during the late stages of crystallization which leaves H_2O-rich fluids that readily dissolve high concentrations of alkalies and silica. Thus, most pegmatites are similar to granites and contain the minerals alkali feldspar and quartz. But other chemical constituents that become concentrated in the residual liquid, like B, Be, and Li are sometimes enriched pegmatites. This leads to crystallization of minerals that are somewhat more rare, such as:

- Tourmaline, $[(Na, Ca)(Mg, Fe, Mn, Li, Al)_3(Al,Fe^{+3})_6Si_6O_{18}(BO_3)_2(OH)_4]$, beryl $[Be_3Al_2Si_6O_{18}]$,
- Lepidolite $[K_2(Li, Al)_{5-6}Si_{6-7}Al_{2-1}(OH,F)_4$,
- Spodumene $[LiAl_2Si_2O_6]$,

which are sometimes found.

Continental Rhyolites

Rhyolites are much more common and voluminous on the continents than in the ocean basins. They range from small domes and lava flows to much larger centers that have erupted volumes measured in 100s of km³ and emplaced as pyroclastic material. Most of the preserved volume is represented as pyroclastic flow deposits, often termed "ash flow tuffs" or "ignimbrites. Large quantities of these deposits were erupted during the middle Tertiary in the western United States, northern Mexico, throughout Central America, and on the western slopes of the Andes mountains. The composition of these deposits is usually metaluminous although peralkaline varieties are known. None are peraluminous in composition. Although the recent examples occur near continental margins. Most seem to be associated with episodes of continental extension, such as in Basin and Range Province of the Western U.S. and Mexico.

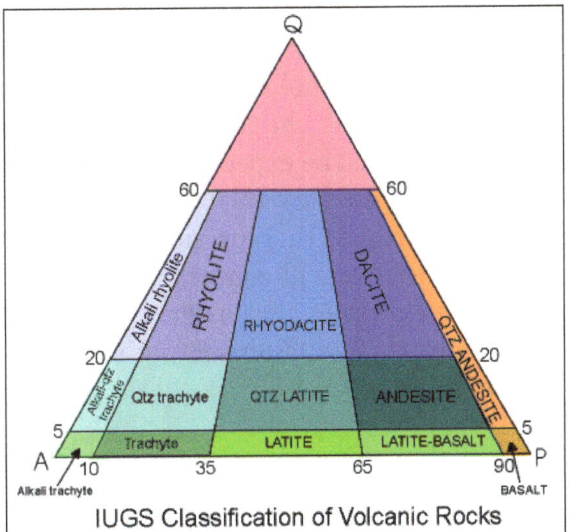

IUGS Classification of Volcanic Rocks

Continental Flood Basalts

Like the large submarine plateaus have erupted on the continents at various times in Earth history.

Province	Age	Original Area Covered (km²)	Types of Basalts %		
Qtz -Thol.	Oliv. Thol.	Alk. Bas.			
Lake Superior	Precambrian	125,000	42	51	7
Siberia	Permo-Triassic	2,500,000	28	69	3
Karoo, S. Africa	Jurassic	2,000,000	57	37	6
Paraná, Brazil	Cretaceous	2,000,000	72	28	0
Deccan, India	Eocene	500,000	55	35	10
Columbia River	Mid-Miocene	163,000	30	70	-

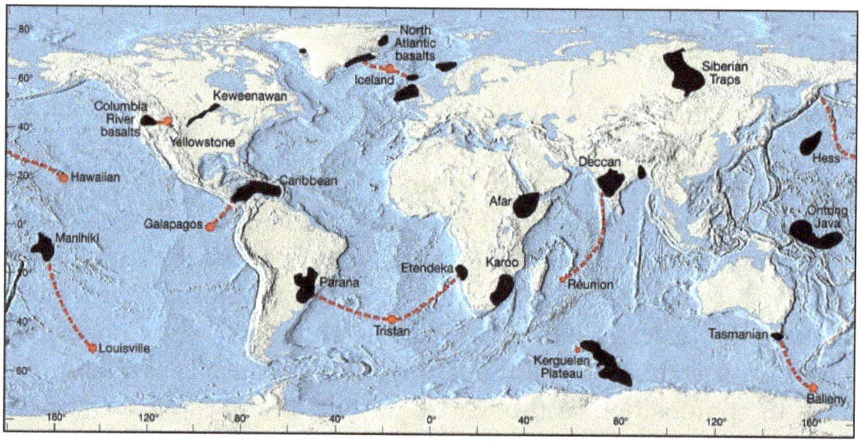

Chemical Composition

Although each flood basalt province differs somewhat in the composition of magmas erupted, most provinces have erupted tholeiitic basalts. With the exception of a few early erupted picrites in some provinces, the tholeiitic basalts tend to have lower concentrations of MgO (5 - 8%) than would be expected from melts that have come directly from the mantle without having suffered crystal fractionation. Thus, despite their large volume, they are differentiated magmas that are similar in many respects to MORBs. Still, they show incompatible trace element concentrations more similar to EMORBs, and have $^{86}Sr/^{87}Sr$ and $^{143}Nd/^{144}Nd$ ratios that extend from the OIB field toward and overlapping with continental crust. This latter feature indicates that they have likely suffered some crustal contamination.

Continental Rift Valleys

Continental Rift valleys are linear zones of extension within continental crust. Some of these extensional zones may be eventually become zones along which the continents break apart to form a new ocean basin, however, there are many examples where such break-ups have failed. The East African Rift which extends from Syria in the north to Mozambique in the south has been active throughout the Cenozoic. During the initial stages of rifting fissure eruptions produced large volumes of basalt and siliceous ignimbrites. During the late Miocene and Pliocene these eruptions became more focused, and produced shield volcanoes consisting of basanites, rhyolites and phonolites. In Plio-Pleistocene times rhyolites were erupted along the main axis of the rift, while basalts continued to be erupted on the plateaus adjacent to the rift. Quaternary volcanoes along the axis of the central rift zones, in Kenya and Tanzania consist of phonolite, trachyte, or peralkaline rhyolite. This province illustrates the wide variety of unusual rock types found in continental rifting settings. Note, however, that parts of the rift along the Red Sea and Gulf of Aden have evolved to oceanic ridges and produce MORBs to form new seafloor.

Coastal Igneous Rocks

The convergent plate margins are the most intense areas of active magmatism above sea level at the present time. Most of world's violent volcanic activity occurs along these zones. In addition, much magmatism also has resulted in (and probably is resulting at present) significant additions to the crust in the form of plutonic igneous rocks.

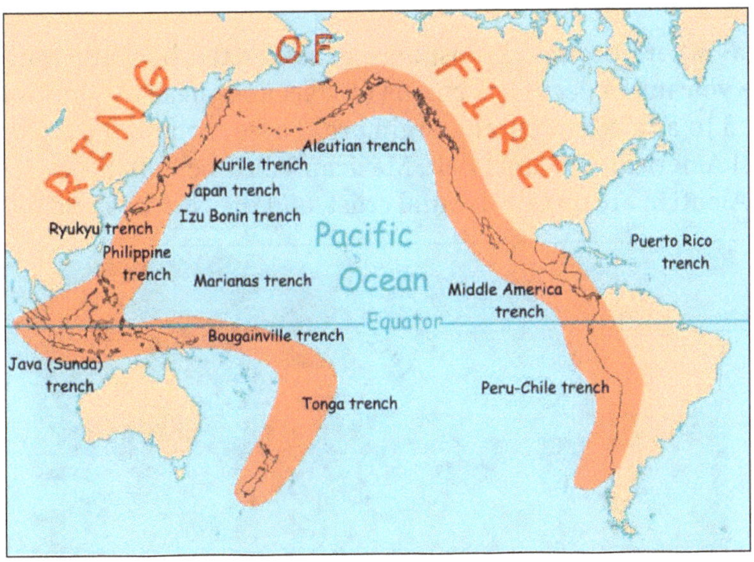

The "Pacific Ring of Fire" surrounds the Pacific Ocean basin and extends into the Indian Ocean and Caribbean Sea. Active subduction is taking place, along these convergent plate boundaries, as evidenced by the zone of earthquakes, called a Benioff Zone that begins near the oceanic trenches and extends to deeper levels in the direction of plate motion. Earthquake focal depths reach a maximum of about 700 km in some areas. Volcanism occurs on the upper plate about 100 to 200 km above the Benioff Zone. For this reason, volcanism in these areas is often referred to as subduction-related volcanism.

Two situations occur:

- In areas where oceanic lithosphere is subducted beneath oceanic lithosphere the volcanism is expressed on the surface as chains of islands referred to as island arcs. These include the Caribbean Arc, the Aleutian Arc, the Kurile Kamachatka Arc, Japan, the Philippines, the South Sandwich Arc, The Indonesian Arc, the Marianas, Fiji, and Solomon Islands.

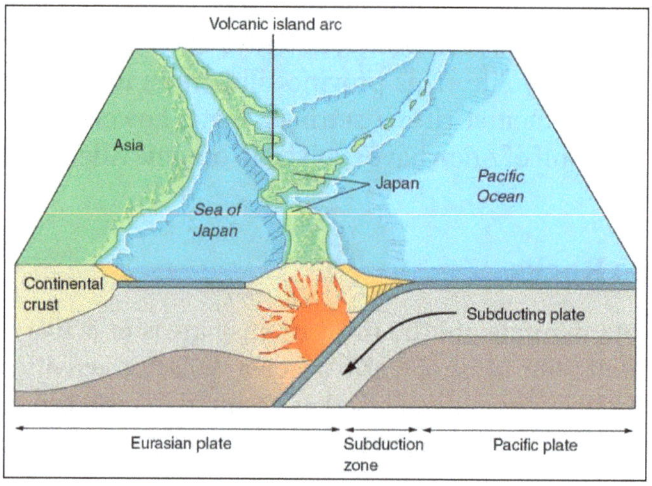

- In areas where oceanic lithosphere is subducted beneath continental lithosphere volcanism occurs as chains of volcanoes near the continental margin, referred to as a continental margin arc. These include the Andes Mountains, Central American Volcanic Belt, Mexican Volcanic Belt, the Cascades, the part of the Aleutian arc on Continental crust, and the North Island of New Zealand.

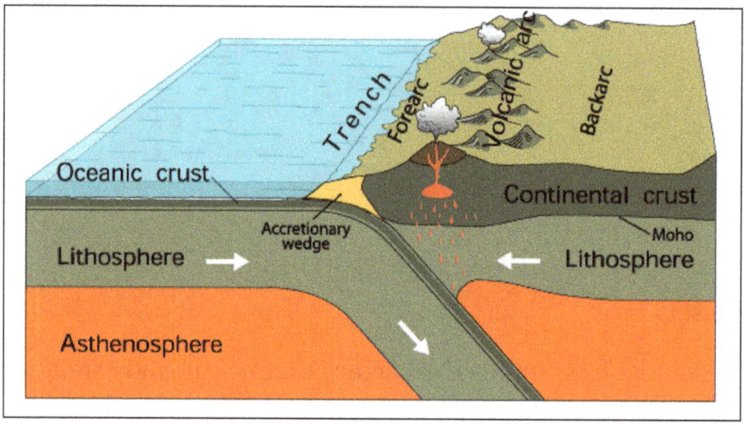

Within these volcanic arcs the most imposing, and therefore most recognized by early workers, features of the landscape are large stratovolcanoes. These usually consist of predominantly andesitic lava flows and interbedded pyroclastic material. But, in the late stages of volcanism more silicic lavas and pyroclastics like dacites and rhyolites are common.

These caldera forming events are usually associated with explosive eruptions that emit silicic pyroclastic material in large-volume eruptions. It is the sudden evacuation of underlying magma chambers that appears to result in the collapse of the volcanoes to form the calderas.

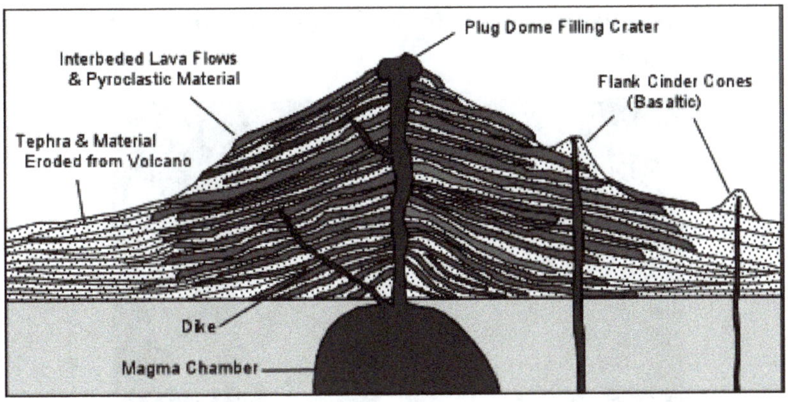

Many of these stratovolcanoes pass through a stage where their upper portions collapse downward to form a caldera.

The imposing presence of these large mostly andesitic stratovolcanoes led to an early widespread perception among petrologists that basalts were rare or absent in these environments.In recent years, however, it has become more evident that basalts are widespread, but do not commonly erupt from the stratovolcanoes.

Instead, they are found in areas surrounding the stratovolcanoes where they erupt to form cinder cones and associated lava flows. One explanation for this distribution is that the magma chambers underlying the stratovolcanoes intercept the basaltic magmas before they reach the surface and allow the basalts to differentiate to more siliceous compositions before they are erupted. Basaltic magmas that are not intercepted by the magma chambers can make it to the surface to erupt in the surrounding areas.

Petrography

Probably the most distinguishing feature of subduction-related volcanic rocks is their usually porphyritic nature, usually showing glomeroporphyritic clusters of phenocrysts. (A textural term used to describe igneous rocks that contain clusters of phenocrysts, which are large crystals in a finer-grained matrix or groundmass).

Basalts commonly contain phenocrysts of olivine, augite, and plagioclase. Andesites and dacites commonly have phenocrysts of plagioclase, augite, and hypersthene, and some contain hornblende.

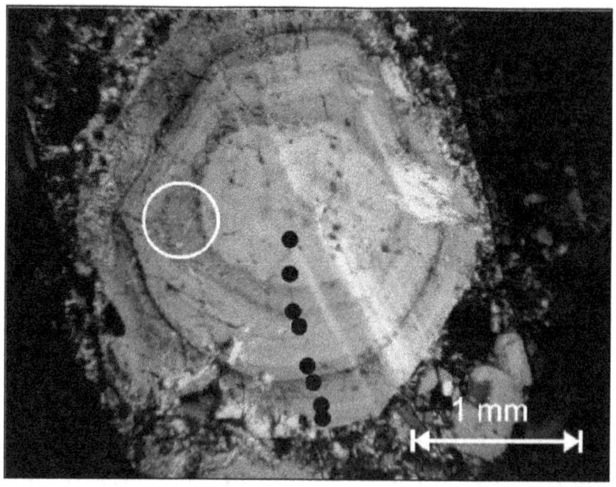

The most characteristic feature of the andesites and dacites is the predominance of fairly calcic plagioclase phenocrysts that show complex oscillatory zoning. Such zoning has been ascribed to various factors, including:

- Kinetic factors during crystal growth: As one zone of the crystal is precipitated the liquid immediately surrounding the crystal becomes depleted in the components necessary for further growth of the same composition. So, a new composition is precipitated until diffusion has had time to renourish the surrounding liquid in the components necessary for the equilibrium composition to form.
- Cycling through a chemically zoned magma chamber during convection: As crystals grow, they are carried in convection cells to warmer and cooler parts of the magma chamber. Some zones are partially dissolved and new compositions are precipitated that are more in equilibrium with the chemical compositions, pressures, and temperatures present in the part of the magma chamber into which the crystal is transported.
- Magma mixing: As magmas mix the chemical compositions of liquids and temperatures change during the mixing process. This could result in dissolution of some zones, and precipitation of zones with varying chemical composition.

Rhyolites occur as both obsidians and as porphyritic lavas and pyroclastics. Phenocrysts present in rhyolites include plagioclase, sanidine, quartz, orthopyroxene, hornblende, and biotite. In addition to these features, petrographic evidence for magma mixing is sometimes present in the rocks, including disequilibrium mineral assemblages, reversed zoning etc. Xenoliths of crustal rocks are also sometimes found, particularly in continental margin arcs, suggesting that assimilation or partial assimilation of the crust could be an important process in this environment.

Possible explanation Calc-alkaline Suite of rocks that are commonly associated with subduction:

- Subduction carries oceanic crust and sediment to depth. As the pressure and temperature rise, the MORB crust and sediment undergo metamorphism that releases hydrous fluids.
- These hydrous fluids carry with them high concentrations of LILE and REE, but leave behind the relatively insoluble HFSE. They also carry the isotopic signature of the basaltic crust sediment mixture that released the fluids, and thus have higher 87Sr/86Sr ratios and lower 143Nd/144Nd ratios, reflecting the isotopic composition of the subducted material. If the sediments are young, they may contribute 10Be to these fluids.
- The fluids act to metasomatize the overlying mantle wedge, enriching it in LILE, REE, B, 87Sr/86Sr, and possibly 10Be and lowering the 143Nd/144Nd ratio of this mantle.
- Adding H_2O to the mantle wedge lowers the solidus temperature allowing for partial melting of this metasomatized mantle and generating hydrous basaltic magmas.
- This hydrous basaltic magmas become saturated with water at crustal depths and differentiate by crystal fractionation, possibly accompanied by contamination of crustal material, to generate the andesites, dacites, and rhyolites of the calc-alkaline suite.

Oceanic Igneous Rocks

The ocean basins cover the largest area of the Earth's surface.Because of plate tectonics, however, most oceanic lithosphere eventually is subducted and thus the only existing oceanic lithosphere is younger than about Jurassic in age and occurs at locations farthest from the oceanic spreading centers.Except in areas where magmatism is intense enough to build volcanic structures above sea level, most of the oceanic magmatism is difficult to access.Samples of rocks can be obtained from drilling, dredging, and expeditions of small submarines to the ocean floor.Numerous samples have been recovered and studied using these methods.Most of the magmatism is basaltic. Still, few drilling expeditions have penetrated through the sediment cover and into the oceanic lithosphere. Nevertheless, we have a fairly good understanding of the structure of the oceanic lithosphere from seismic studies and ophiolites.

Ophiolites

An ophiolite is a sequence of rocks that appears to represent a setion through oceanic crust. Ophiolites occur in areas where obduction (the opposite of subduction) has pushed a section of oceanic lithosphere onto continental crust. During this process, most ophiolite sequences have been highly deformed and hydrothermally altered. Nevertheless, it is often possible to look through the deformation and alteration and learn something of the structure of oceanic lithosphere. An idealized ophiolite sequence shows an upper layer consisting of deep sea sediments (limestones, cherts, and shales), overlying a layer of pillow basalts. Pillow basalts have a structure consisting of overlapping pillow-shaped pods of basalt. Such pillow structure is typical of lavas erupted under water. The pillow basalts overly a layer consisting of numerous dikes, some of which were feeder dikes for the overlying basalts. Beneath the sheeted dike complex are gabbros that likely represent the magma chambers for the basalts. The upper gabbros are massive while the lower gabbros show layering that might have resulted from crystal settling.

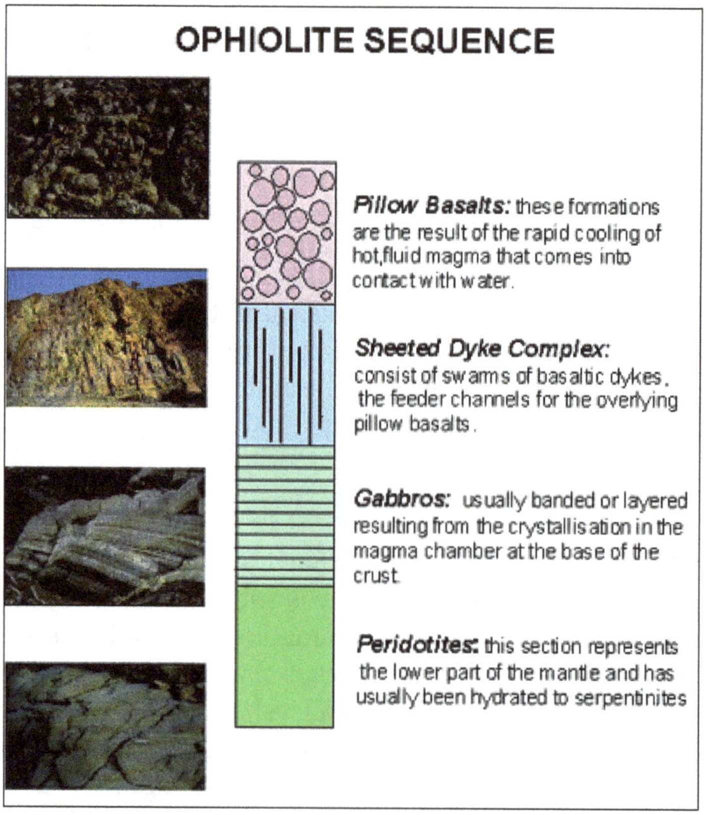

At the base of the layered gabbros, there is a sharp increase in the density of the rocks, and the composition changes to ultramafic rocks. This sharp change in density is correlated with what would be expected at the base of the crust, and is thus referred to as the petrologic moho. At the top of the ultramafic sequence the rock type is harzburgite (Ol + Opx), a rock type expected to be the residual left from partially melting peridotite. The base of the ultramafic layer is composed of peridotite. Because most ophiolites have been hydrothermally altered, most of the mafic rocks have been altered to serpentinite. Note that ophiolite means "snake rock".

Volcanic Settings

Volcanism occurs at three different settings on the ocean floor.

- Oceanic Ridges: These are the oceanic spreading centers where a relatively small range of chemical compositions of basalts are erupted to form the basaltic layer of the oceanic crust. This chemical type of basalt is referred to as Mid Ocean Ridge Basalts (MORBs). In some areas, particularly Iceland, where there has been a large outpouring of basalts on the oceanic ridge, basalts called Enriched Mid Ocean Ridge Basalts (EMORBs)have been erupted.
- Oceanic Islands: These are islands in the ocean basins that generally occur away from plate boundaries, and are often associated with hot spots,. A wide variety of rocks occur in these islands, not all are basaltic, but those that aren't appear to be related to the basaltic magmas. In general these rocks are referred to as Oceanic Island Basalts (OIBs).
- Large Igneous Provinces (LIPs): These are massive outpourings of mostly basaltic lavas that have built large submarine plateaus. Most are mid-Cretaceous in age. They are not well studied, but most have compositions similar to OIBs, and some may have once had oceanic islands on top, but most of these have since been removed by erosion.

Basalts

On a chemical basis, basalts can be classified into three broad groups based on the degree of silica saturation. This is best seen by first casting the analyses into molecular CIPW norms (the same thing as CIPW norms except the results are converted to mole % rather than weight %). On this basis, most basalts consist predominantly of the normative minerals - Olivine, Clinopyroxene, Plagioclase, and Quartz or Nepheline. These minerals are in the 4 component normative system Ol-Ne-Cpx-Qtz, shown here as a tetrahedron. In the tetrahedron, plagioclase plots between Ne and Qtz, and Opx plots between Ol and Qtz. The basalt tetrahedron can be divided:

- The plane Cpx-Plag-Opx is the critical plane of silica saturation.Compositions that contain Qtz in their norms plot in the volume Cpx-Plag- Opx-Qtz, and would be considered silica oversaturated.Basalts that plot in this volume are called Quartz Tholeiites.
- The plane Ol - Plag - Cpx is the critical plane of silica undersaturation. Normative compositions in the volume between the critical planes of silica undersaturation and silica saturation are silica saturated compositions (the volume Ol - Plag - Cpx - Opx). Silica saturated basalts are called Olivine Tholeiites.
- Normative compositions that contain no Qtz or Opx, but contain Ne are silica

undersaturated (the volume Ne-Plag-Cpx-Ol). Alkali Basalts, Basanites, Nephelinites, and other silica undersaturated compositions lie in the silica undersaturated volume.

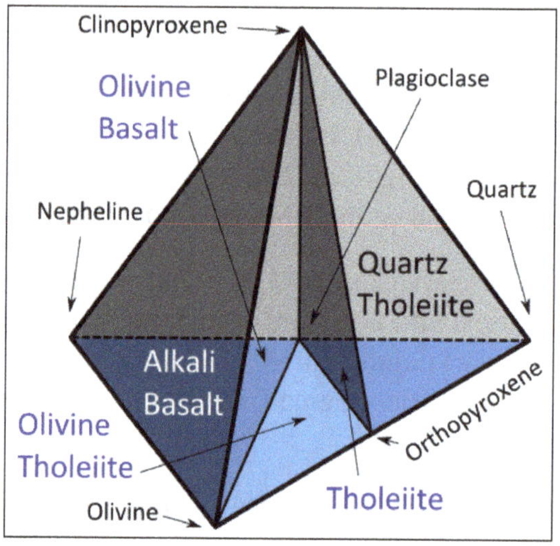

Note that tholeiitic basalts are basalts that show a reaction relationship of olivine to liquid which produces a low-Ca pyroxene like pigeonite or Opx.Both olivine tholeiites and quartz tholeiites would show such a relationship and would eventually precipitate either Opx or pigeonite. The critical plane of silica undersaturation appears to be a thermal divide at low pressure. This means that compositions on either side of the plane cannot produce liquids on the other side of the plane by crystal fractionation. To see this, look at the front two faces of the basalt tetrahedron. These are in the three component systems Ol-Cpx-Qtz and Ol-Cpx-Ne.

Mid Ocean Ridge Basalts (MORBs)

Occurrence

The Oceanic Ridges are probably the largest producers of magma on Earth. Yet, much of this magmatism goes unnoticed because, with the exception of Iceland, it all takes place below the oceans. This magmatism is responsible for producing oceanic crust at divergent plate boundaries. Magma is both erupted and intruded near the central depressions that form the oceanic ridges. Thus, both basalts and gabbros are produced. But, little is known of the gabbros since they are rarely exposed and most oceanic lithosphere eventually is subducted. The main melting mechanism is likely decompression melting as rising convection cells move upward through the mantle beneath the ridges. At most oceanic ridges the basalts that are erupted are tholeiitic basalts sometimes referred to as NMORBs (normal MORBs).

At the oceanic ridges, the basalts erupted range in composition from Olivine tholeiites to Quartz tholeiites. The compositions are by and large restricted to basalt, i.e. less than about 52% SiO_2. The diagram shown here is called an AFM diagram. It is a triangular variation diagram that plots total alkalies at the A corner, total iron at the F corner, and MgO at the M corner. MORBs show a restricted range of compositions that fall along a linear trend extending away from the compositions of Mg-rich pyroxenes and olivines.

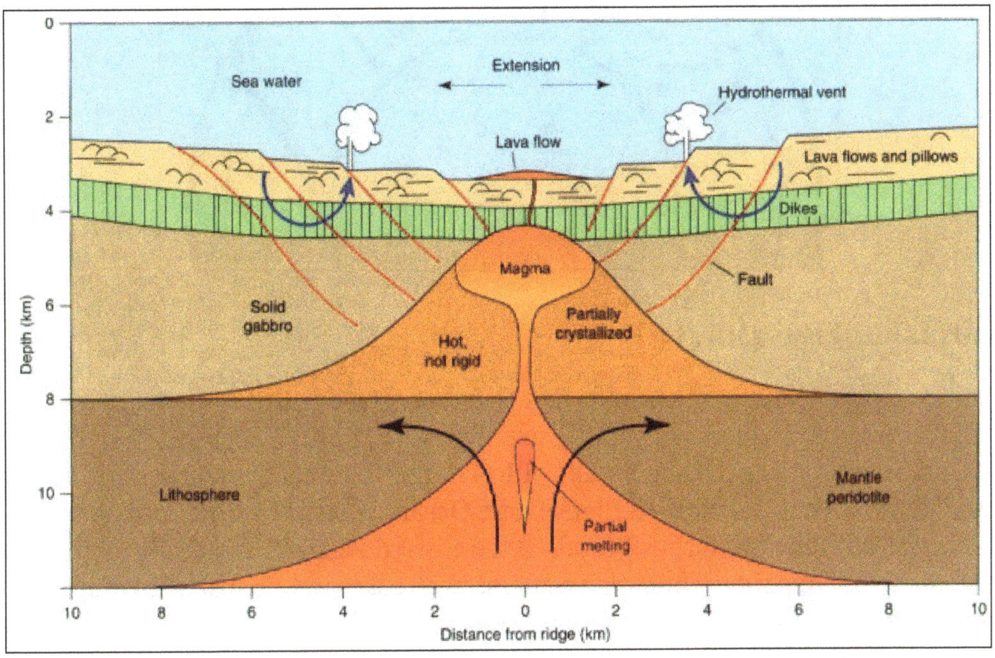

This is the trend that would expected from fractional crystallization involving the removal of early crystallizing olivines and pyroxenes from a tholeiitic basaltic liquid. Note that the trend is often referred to as an Fe-enrichment trend.

Some thoughts on the origin of MORBs:

- Melting is likely caused by decompression of the mantle as it rises beneath the oceanic ridges as a result of convection.
- Primary MORB magmas appear to be produced by partial melting of the mantle at pressures between 15 and 20 kb.
- Most MORBs erupted are not primary melts of the mantle, but instead appear to have suffered olivine fractionation.
- The small range in composition of MORBs can be explained by crystal fractionation of Olivine + Plagioclase + Cpx at low pressures near the surface.
- MORBs appear to be the result of melting of an incompatible element depleted mantle, both in terms of their incompatible trace element compositions and isotopic ratios of Sr and Nd.

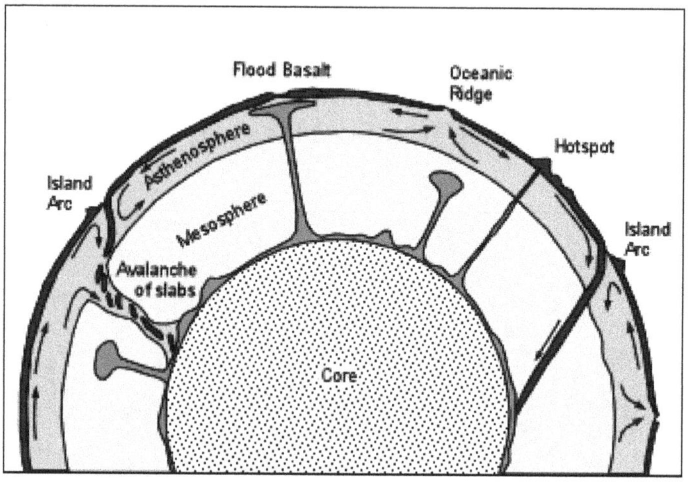

Ocean Island Basalts (OIBs)

Oceanic islands are, in general, islands that do not occur along the divergent or convergent plate boundaries in the ocean basins. Nevertheless, EMORBs, such as those that occur in Iceland, as well as the Alkalic basalts of Iceland have much in common with magmas erupted in the oceanic islands. In the Atlantic Ocean, which is a slow-spreading oceanic basin, as well as in the Galapagos Islands of the eastern Pacific Ocean, some of the islands occur close to oceanic ridge spreading centers. In all cases we must keep in mind that the parts of these islands that are accessible for sampling represent only a fraction of the mass of the volcanic structures which rise from the ocean floor at depths up to 10,000 m. Thus as with the ocean ridge volcanic rocks, there is a potential sampling problem.

Oceanic Islands

Most oceanic islands appear to be related to ascending plumes of hot mantle. These plumes must be relative narrow features because they appear to operate independent of the main convection cells that ascend beneath the oceanic ridges and descend at subduction zones. Still, in places like Iceland on the ocean ridge, magma production rates are high, and compositions of rocks are similar to those found in oceanic islands. So Iceland could also be considered an oceanic island. If these rising plumes of hot mantle remain stationary in their positions in the mantle, they produce hot spots, as discussed previously. Hot spots are most recognizable when they occur beneath plates that move with higher velocities. Beneath faster moving plates, like the Pacific Plate, this results in linear chains of islands.

At the position directly over the hotspot, rising mantle melts to produce magma that erupts on the seafloor, eventually building a volcanic island directly over the hot spot. As the lithospheric plate moves over the hot spot the volcano eventually is cut off from its source of magma, and becomes extinct, and a new volcano forms on the plate at the location directly above the hot spot.

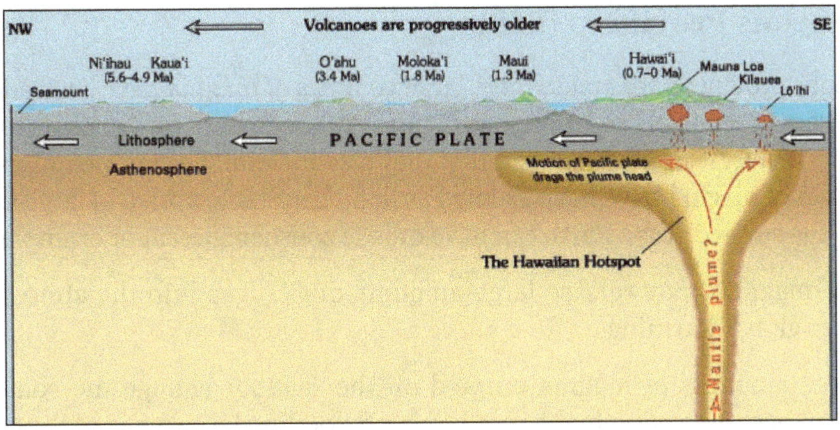

The volcanoes that have moved away from the hot spot eventually begin to erode until their elevations are reduced below sea level. At this point they are called seamounts.

Such linear chains of islands and sea mounts are most evident in the Pacific ocean. The largest of these is the Hawaiian - Emperor chain. The hot spot that produced this chain is currently located under the position of the big island of Hawaii, which has the only currently active volcanoes in the chain. The bend in the Hawaiian-Emperor chain must have resulted from a change in the direction of plate motion. Volcanic rocks dredged from the sea floor at the location of this bend are about 40 million years old. Thus, prior to 40 million years ago the Pacific Plate was moving in a more northerly direction. The most northerly seamount is dated at about 60 million years. Seamounts older than 60 million have apparently been subducted.

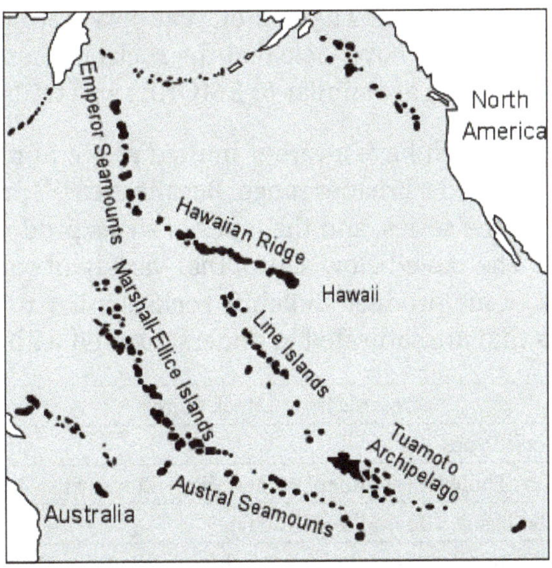

The reason such island/seamount chains are not as evident in the other oceans is because the plate velocity is lower and volcanoes tend to remain over the hot spots for longer periods of time, building elongated groups of islands rather than linear chains.

Large Igneous Provinces (LIP)s

Large igneous provinces are areas where large volumes of magma have been added to the Earth's crust over relatively short periods of time. Although here we discuss these in terms of the ocean basins, it should be noted that they also include the continental areas where large volumes of magma have been erupted as flood basalts. Eruption of large amounts of magma on the surface of the Earth can have drastic consequences. For example:

- The magmas may release large amounts of CO_2 gas into the atmosphere and force global warming.
- Large amounts of magma erupted on the seafloor change the volume of the ocean basins causing higher sea level and flooding large areas of the continents.

The mid Cretaceous Period was a time of higher than normal global temperatures and high stands of the oceans. Eruption of magma on the ocean floor at this time might have been the cause of these conditions. Evidence is preserved on the sea floor in the form of large submarine plateaus that were emplaced during this time period.

In the Pacific Ocean, much of the oceanic lithosphere east of the position of the ridge 80 million years ago has been subducted. Thus, if the submarine plateaus formed at the ridge, then it would be expected that half of each plateau became separated at the ridge and have since been subducted. These probable plateaus are shown in the map, and if they were present would double the original size of the plateau. Thus, for example the largest of the plateaus is the Ontong-Java Plateau, now located in the southeastern Pacific, with a volume of about 50 million km^3.But if the other half had been present, the total volume of magma erupted over a 25 million year period would have been over 100 million km^{3}The few studies that have looked at the rocks in these submarine plateaus suggest that their compositions are similar to EMORBs and OIBs.

Unlike the ocean ridges, which have a rather limited range of rock compositions, the oceanic islands have produced a broader range. Basalts are still predominant, but other compositions are part of the series, and the types of rocks produced are variable from one island to the next. The table below shows that variety of rock types found at different oceanic islands. Some produce tholeiitic rocks similar to EMORBs and others produce alkalic basalts that are saturated to undersaturated with respect to silica.

Oceanic Island Rock Suites	
Island or Group	Rock Types
Ascension	Oliv. Tholeiite (dominant) + Hawaiite + Mugearite + Trachyte + Peralk. Rhyolite
Azores	Alk. basalt + Hawaiite + Trachyte
Fernando de Noronha	Alk. Basalt + Nephelinite + Trachyte + Alkali Basalt + Trachyte + Phonolite
St. Helena	Alk. Basalt + Mugearite + Hawaiite + Trachyte + Phonolite
Trinadade	Nephelinite + Phonolite (dominant)
Tristan de Cunha	Alk. Basalt + Trachybasalt (dominant) + Trachyte

Gough	Alk. Basalt + Ol Tholeiite + Hawaiite + Trachyte
Réunion	Ol Tholeiite (dominant) + Mugearite
Mauritius	Alk. Basalt (dominant) + Mugearite + Phonolitic Trachyte
Hawaii	Tholeiite (dominant) + Alkali Basalt + Hawaiite + Mugearite + Trachyte
Tahiti	Alk. Basalt + Mugearite + Hawaiite + Trachyte
Galapagos	Tholeiite + Alk. Basalt + Icelandite (minor) + Qtz Trachyte (minor)
Jan Mayen	Alk. Basalt (dominant) + Trachyte

Origin of OBIs

- Melting is caused by decompression of the mantle as it rises in narrow plumes beneath the oceanic islands.
- Melting probably takes place at relatively low pressure to produce silica saturated tholeiitic magmas, but at high pressure to produce the silica undersaturated varieties such alkali basalts, basanites and nephelinites.
- These basic magmas evolve by crystal fractionation to produce a wide variety of other rock types including hawaiites, mugearites, trachytes, and rhyolites. The more silica undersaturated basanites would fractionate to produce nephelinites and phonolites.
- Lower degrees of melting produce the most incompatible element rich magmas, like basanites, while successively higher degrees of melting could produce the alkali basalts and tholeiitic basalts.
- OIBs (including EMORBs) appear to be the result of melting of an incompatible element enriched mantle, both in terms of their incompatible trace element compositions and isotopic ratios of Sr and Nd.

Igneous Rock Classification

The main QAPF classification for plutonic and volcanic rocks which is based on the modal mineral proportions of quartz (Q), alkali feldspar (A) and plagioclase (P) or of alkali feldspar (A), plagioclase (P) and feldspathoids (F). Rocks with mafic content >90% have their own classification. This is followed by the classification separates and individually classifies the pyroclastic, carbonatitic, melititic, lamprophyric and charnockitic rocks before entering If the mineral mode cannot be determined as is often the case for volcanic rocks, then a chemical classification of total alkalis versus silica (TAS) is used. The nomenclature for these classifiations necessitates only 297 rock names out of the about 1500 that exist.

Ten Principles of IUGS Igneous Rock Classification

1. Use descriptive attributes.
2. Use actual properties.

3. Ensure suitability for all geologists.
4. Use current terminology.
5. Define boundaries of rock species.
6. Keep it simple to apply.
7. Follow natural relations.
8. Use modal mineralogy.
9. If mode not feasible, use chemistry.
10. Follow terminology of other iugs advisory bodies.

Main QAPF Classification

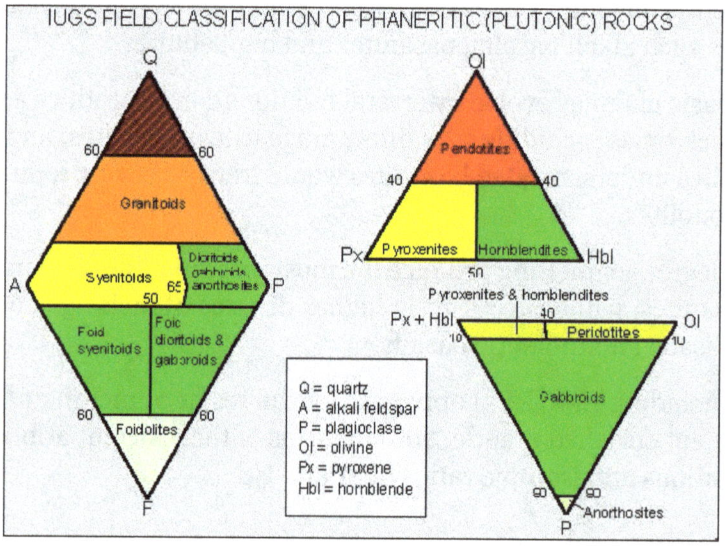

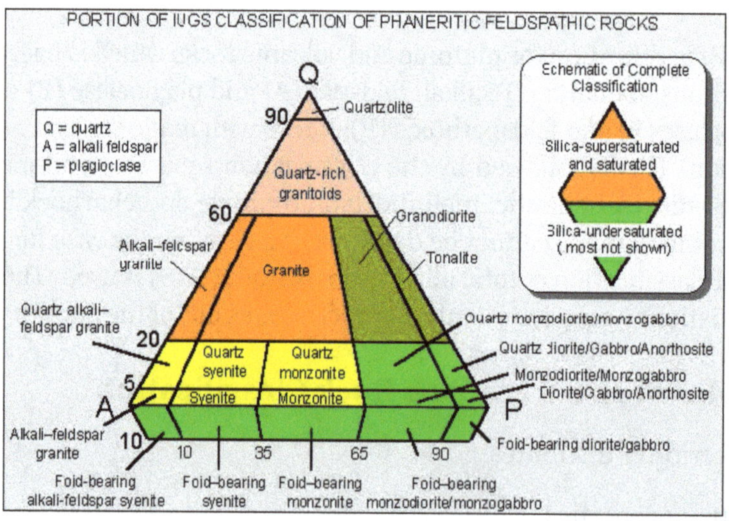

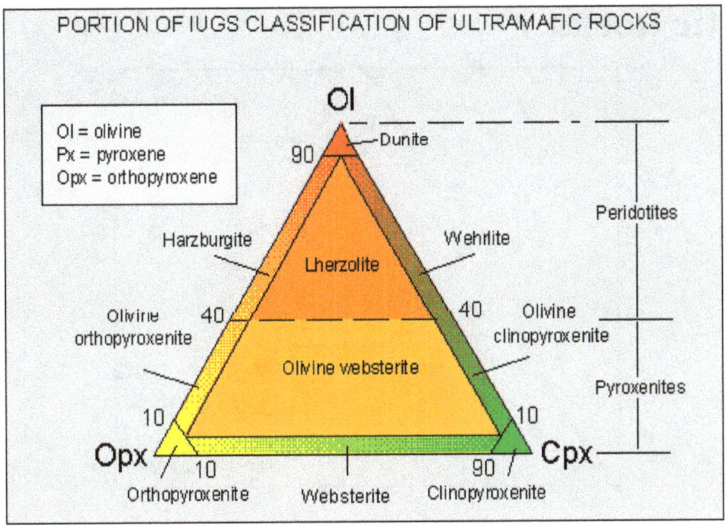

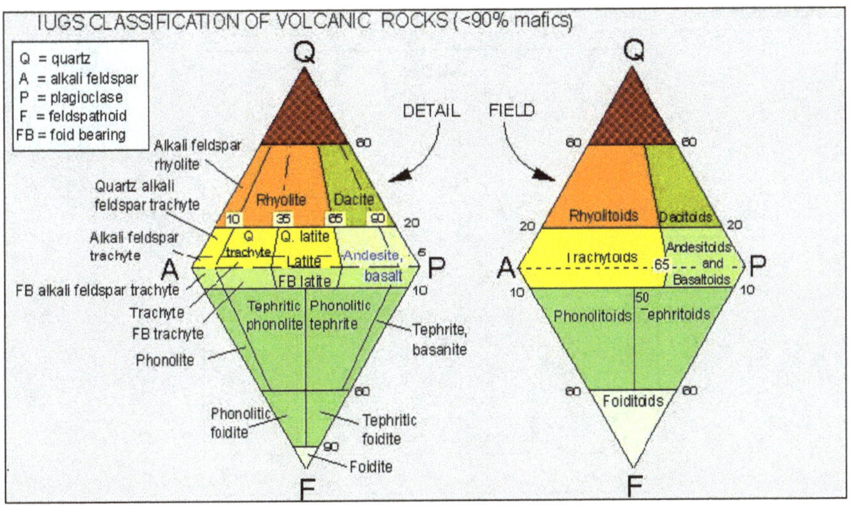

Pyroclastic Rocks

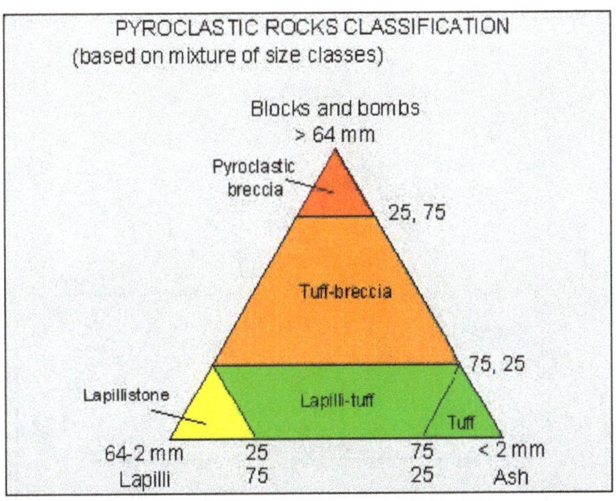

Carbonatitic Rocks

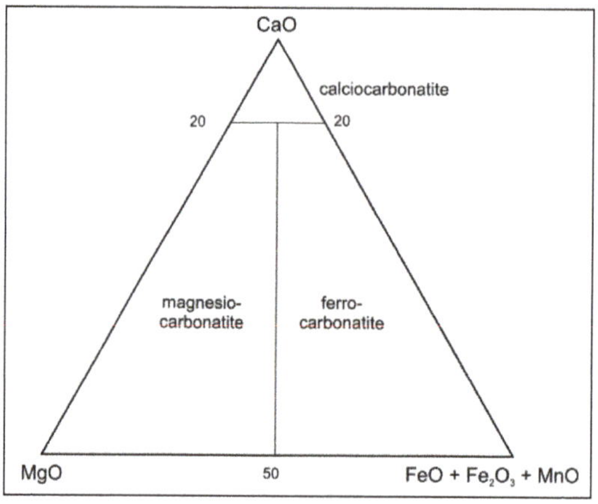

Melititic Rocks

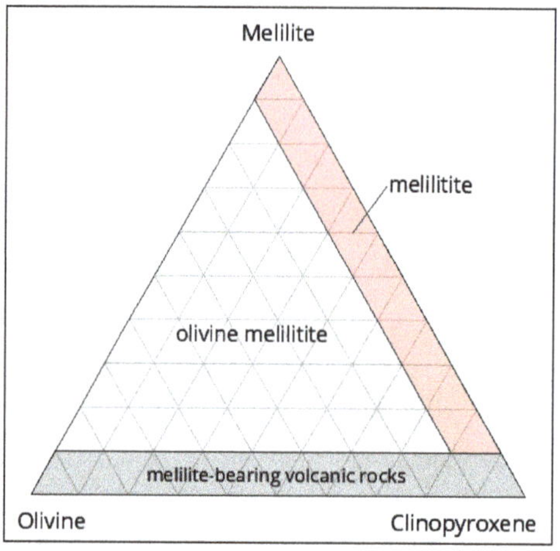

Lamprophyric Rocks

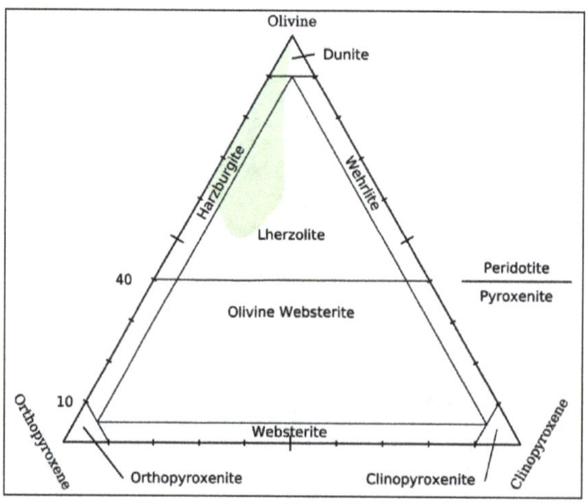

Charnockitic Rocks

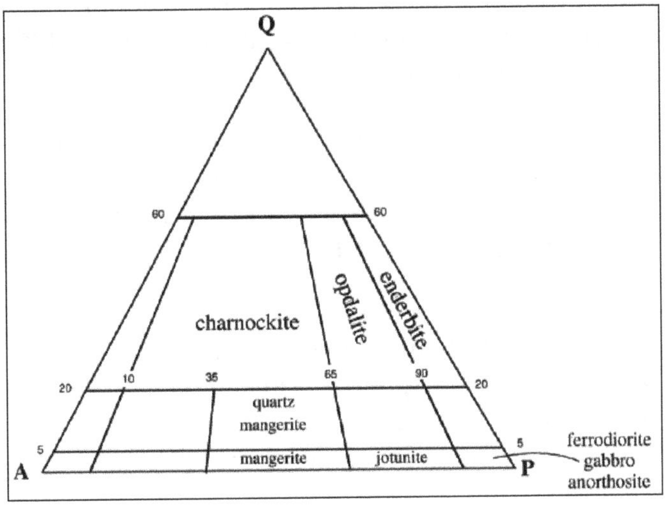

Origins and Differentiation of Magma

Magmas must require special circumstances in order to form and do not form just anywhere in the crust. The different layers of the earth have differing chemical compositions.

- Crust: Variable thickness and composition:
    - Continental 10 - 70 km thick, underlies all continental areas, has an average composition that is andesitic.
    - Oceanic 8 - 10 km thick, underlies all ocean basins, has an average composition that is basaltic.
- Mantle - 3488 km thick, made up of a rock called peridotite (Olivine + Opx + Cpx). Evidence comes from Seismic wave velocities, experiments, and peridotite xenoliths (foreign rocks) brought to the surface by magmas. Experimental evidence suggests that the mineralogy of peridotite changes with depth (or pressure) in the Earth. At low pressure, the mineral assemblage is Olivine + Cpx + Opx + Plagioclase (plagioclase peridotite). At higher pressure the assemblage changes to Olivine + Cpx + Opx + Spinel $[(Mg,Fe^{+2}) (Cr, Al, Fe^{+3})_2O_4]$ (spinel peridotite). At pressures above about 30 kilobars, the assemblage changes to Olivine + Cpx + Opx + garnet (garnet peridotite). This occurs because Al changes its coordination with increasing pressure, and thus new minerals must form to accommodate the Al. At greater depths, such as the 400 km discontinuity and the 670 km discontinuity, olivine and pyroxene likely change to high pressure polymorphs. Despite these changes in mineral assemblage, the chemical composition of the mantle does not appear to change much in terms of its major element composition.

- Core - 2883 km radius, made up of Iron (Fe) and small amount of Nickel (Ni). Evidence comes from seismic wave velocities, experiments, and the composition of iron meteorites, thought to be remnants of other differentiated planets that were broken apart due to collisions.
- Lithosphere: About 100 km thick (up to 200 km thick beneath continents), very brittle, easily fractures at low temperature. Note that the lithosphere is comprised of both crust and part of the upper mantle. The plates that we talk about in plate tectonics are made up of the lithosphere, and appear to float on the underlying asthenosphere.
- Asthenosphere: About 250 km thick - solid rock, but soft and flows easily (ductile). The top of the asthenosphere is called the Low Velocity Zone (LVZ) because the velocities of both P- and S-waves are lower than the in the lithosphere above. But, not that neither P- nor S-wave velocities go to zero, so the LVZ is not completely liquid.
- Mesosphere: About 2500 km thick, solid rock, but still capable of flowing.
- Outer Core: 2250 km thick - liquid. We know this because S-wave velocities are zero in the outer core. If $V_s = 0$, this implies m = 0, and this implies that the material is in a liquid state.
- Inner core: 1230 km radius, solid.

Origins of Magma

Magmas are not likely to come from the only part of the Earth that is in a liquid state, the outer core, because it does not have the right chemical composition. The outer core is made mostly of Fe with some Ni, magmas are silicate liquids.

In the ocean basins, magmas are not likely to come from melting of the oceanic crust, since most magmas erupted in the ocean basins are basaltic. To produce basaltic magmas by melting of the basaltic oceanic crust would require nearly 100% melting, which is not likely. In the continents, both basaltic and rhyolitic magmas are erupted and intruded. Basaltic magmas are not likely to have come from the continental crust, since the average composition is more siliceous, but more siliceous magmas (andesitic - rhyolitic) could come from melting of the continental crust. Basaltic magmas must come from the underlying mantle. Thus, with the exception of the continents, magmas are most likely to originate in the mantle from melting of mantle peridotite.

Temperature varies with depth or pressure in the Earth along the geothermal gradient. The normal geothermal gradient is somewhat higher beneath the oceans than beneath the continents, at least at shallow levels. If we compare the normal geothermal gradients with the experimentally determined phase diagram for peridotite containing little water or carbon dioxide, we find that the peridotite solidus temperature is everywhere higher than the normal geothermal gradients. Thus, under normal conditions the mantle is solid, as we would suspect from the seismic evidence.

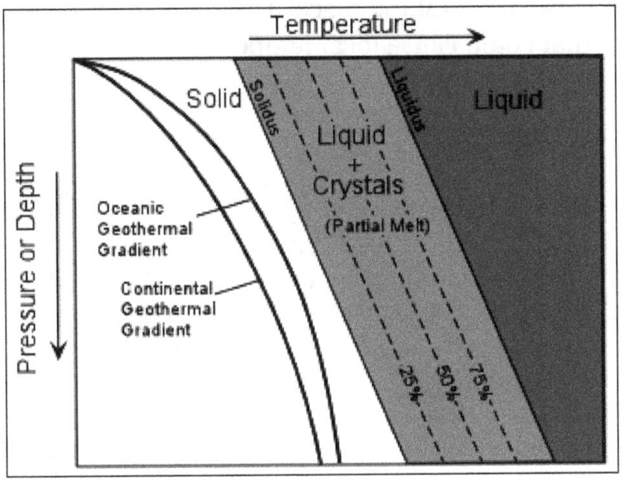

Thus, in order to generate a melt, either we must find a way to increase the geothermal gradient so that it is above the peridotite solidus or reduce the temperature of the peridotite solidus. In either case note that all we have to do is get the temperature in some part of the Earth, as expressed by the geothermal gradient, into the field of partial melt. Partial melting is the most likely case because it requires less of an increase in temperature or less of a decrease in the peridotite solidus. Once a partial melt has formed, the liquid portion can be easily separated from the remaining solids since liquids are more mobile and, in general, have a lower density than solids.

Raising the Geothermal Gradient

Radioactive Heat

Elements like U, Th, K, and Rb have radioactive isotopes. During radioactive decay, sub-atomic particles are released by the decaying isotope and move outward until they collide with other atomic particles. Upon collision, the kinetic energy of the moving particles is converted to heat. If this heat cannot be conducted away, then the temperature will rise. Most the heat within the Earth is generated by radioactive decay, and this is the general reason why temperature increases with depth in the Earth. But most the radioactive isotopes are concentrated in the crust. Although there are areas in the continental crust where high concentrations of radioactive elements have locally raised the temperature, at least high enough to cause metamorphism, this is a rare occurrence. It is even more unlikely that areas of high concentration develop within the mantle. Thus, concentrations of radioactive elements are not likely to cause melting.

Frictional Heat

In areas where rocks slide past one another, such as at the base of the lithosphere, on at subduction zones, heat could be generated by friction. If this heat cannot be conducted away fast enough, then it may cause a localized rise in temperature within the zone where the sliding or shearing is taking place. This could cause a localized spike on the geothermal gradient that could cause local temperatures to rise above the solidus.

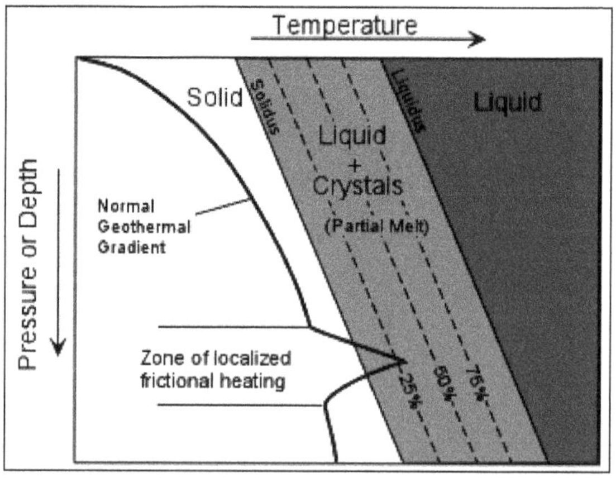

Decompression due to Convection

Convection is a form of heat transfer wherein the heat moves with the material. Convection can be induced if the temperature gradient is high enough that material at depth expands so that its density is lower than the material above it. This is an unstable situation and the hotter, lower density material will rise to be replaced by descending cooler material in a convection cell.

The rate of convection depends on both on the temperature gradient and the viscosity of the material (note that solids convect, but the rate is lower than in liquids because solids have higher viscosity). In the Earth, temperature gradients appear to be high enough and viscosity low enough for convection to occur. Plate tectonics appears to be driven by convection in some form.

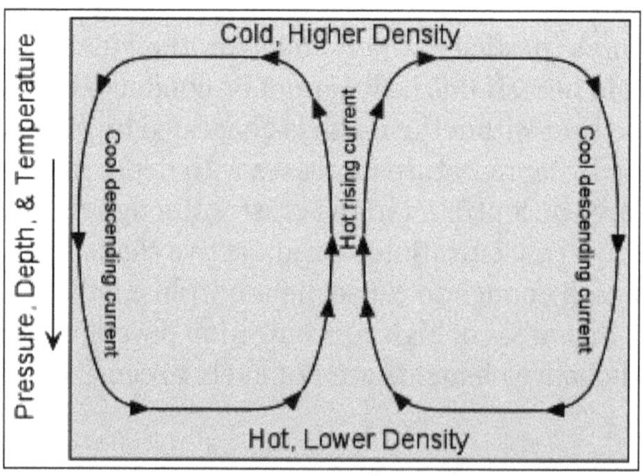

Anywhere there is a rising convection current, hotter material at depth will rise carrying its heat with it. As it rises to lower pressure (decompression) it will cool somewhat, but will still have a temperature higher than its surroundings. Thus, decompression will result in raising the local geothermal gradient. If this new geothermal gradient reaches temperatures greater than the peridotite solidus, partial melting and the generation of magma can occur. This mechanism is referred to as decompression melting.

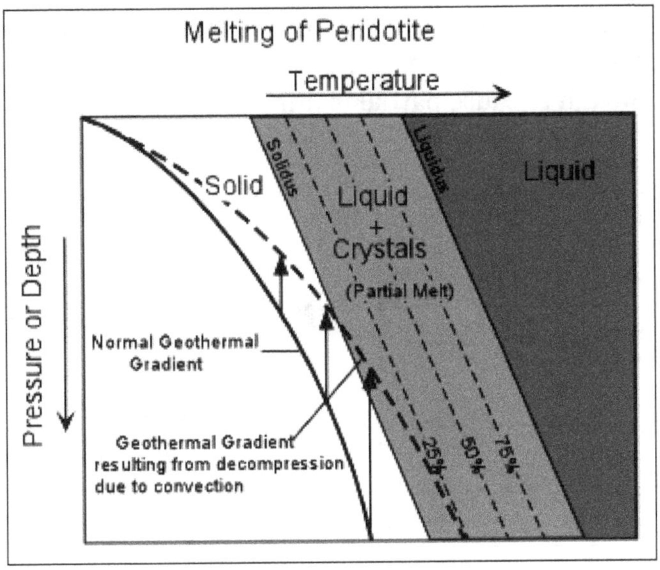

Lowering the Solidus Temperature

Mixtures of components begin melting at a lower temperature than the pure components. In a two component system addition of a third component reduces both the solidus and liquidus temperatures. This suggests that if something can be added to the mantle, it could cause the solidus and liquidus temperatures to be lowered to the extent that the solidus could become lower than the geothermal gradient and result in partial melting, without having to raise the geothermal gradient. Such a melting mechanism is referred to as flux melting.

It's difficult to imagine how solid components could be added to the mantle. But volatile components, for example H_2O and CO_2, because of the high mobility, could be added to the mantle, particularly at subduction zones.

- Oceanic crust is in contact with sea water, thus water could be in oceanic crust both due to weathering, which produces hydrous minerals like clay minerals, and could be in the pore spaces in the rock.
- Oceanic sediments eventually cover the basaltic oceanic crust produced at oceanic ridges. Much of this sediment consists of clay minerals (which contain water) and carbonate minerals (which contain carbon dioxide).
- As the oceanic lithosphere descends into the mantle at a subduction zone, it

will be taken to increasingly higher temperatures as it gets deeper. This will result in metamorphism of both the basalt and the sediment. Metamorphism is essentially a series of dehydration and decarbonation reactions, i.e. chemical reactions that transform hydrous and carbonate minerals into nonhydrous minerals and give up H_2O and CO_2 as a fluid phase.

- Addition of this fluid phase, either to the subducted lithosphere or the mantle overlying the subducted lithosphere could lower the solidus and liquidus temperatures enough to cause partial melting.

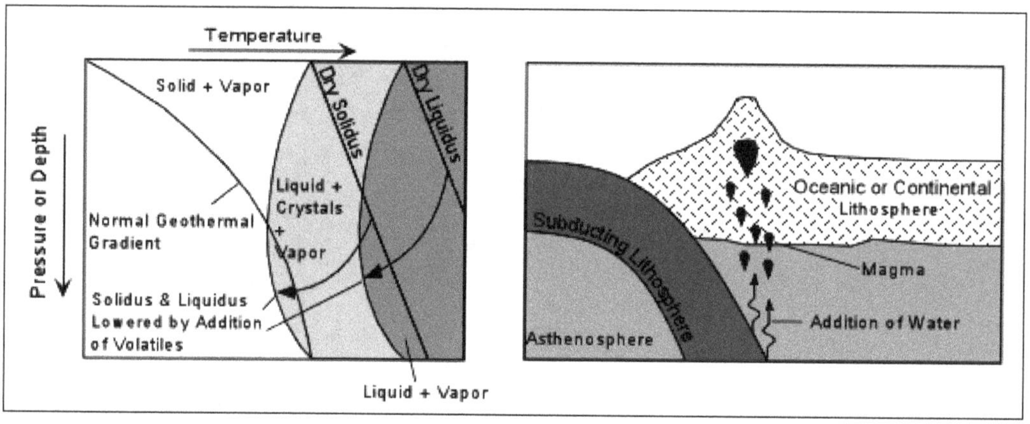

Crustal Anatexis

In the continental crust, it is not expected that the normal geothermal gradient will be high enough to cause melting despite the fact that hydrous and carbonate minerals occur in many continental rocks. Furthermore, because continental rocks are at low temperature and have a very high viscosity, convective decompression is not likely to occur. There is evidence that continental crustal rocks sometimes melt. This is called crustal anatexis. The following scenario is one mechanism by which crustal anatexis could occur:

- Basaltic magmas, generated in the mantle, by flux melting, decompression melting or frictional heat, rise into the crust, carrying heat with them.
- Because basaltic liquids have a higher density than crust, they may not make it all the way to the surface, but instead intrude and cool slowly at depth.
- Upon cooling the basaltic magmas release heat into the crust, raising the geothermal gradient (increasing the local temperature).
- Successive intrusions of mantle-derived mantle into the same area of the crust may cause further increases in temperature, and eventually cause the geothermal gradient to become higher than the wet solidus of the crustal material, resulting in a partial melt of the crust.

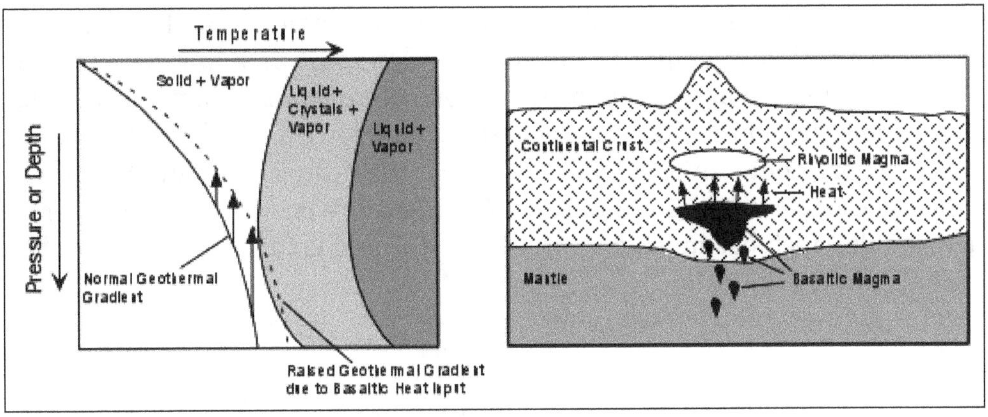

Magmatism and Plate Tectonics

It should be obvious that magmatism is closely related to plate tectonics. The diagram below summarizes melting mechanisms that occur as a result of plate tectonics and may be responsible for the generation of magmas in a variety of plate tectonic settings, such as oceanic ridges, near subduction zones, and at rift valleys.

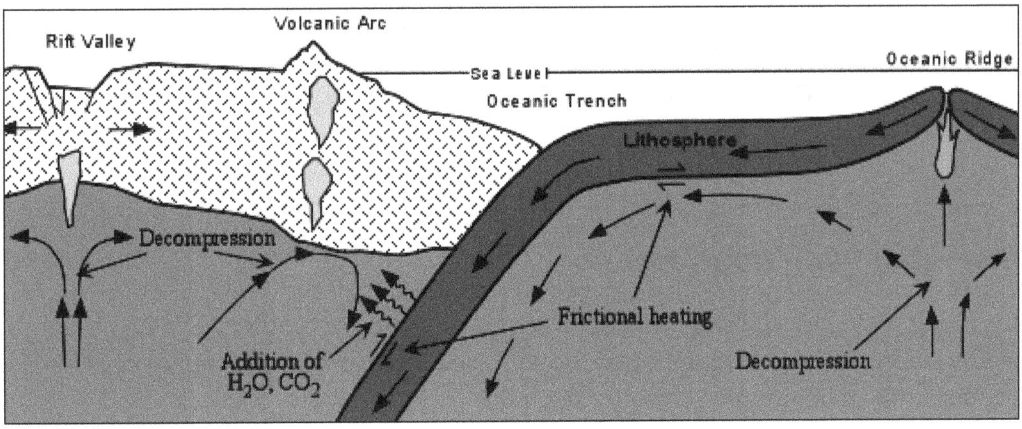

Diverging Plate Boundaries

Diverging plate boundaries are where plates move away from each other. These include oceanic ridges or spreading centers, and rift valleys. Oceanic Ridges are areas where mantle appears to ascend due to rising convection currents. Decompression melting could result, generating magmas that intrude and erupt at the oceanic ridges to create new oceanic crust.

Iceland is one of the few areas where the resulting magmatism has been voluminous enough to built the oceanic ridge above sea level.

Continental Rift Valleys or Extensional Zones are areas, usually located in continental crust where extensional deformation is occurring. These areas may be incipient spreading centers and may eventually evolve into oceanic ridges, such as has occurred in the Red Sea region. Whether or not they develop into spreading centers, they are likely caused by mantle upwelling below the zone of extension. Mantle upwelling may result in decompression melting of the mantle, and could induce crustal anatexis. A good example of a continental rift valley is the East African Rift Valley. Another example is the Rio Grande Rift in Colorado and New Mexico, which is part of a larger region of extension that includes much of the western U.S. and is called the Basin and Range Province (Eastern California].

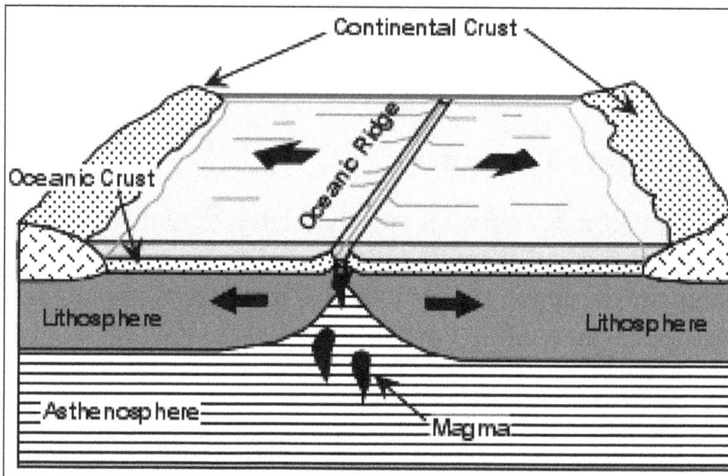

Diverging Plate Boundary
Oceanic Ridge - Spreading Center

Converging Plate Boundaries

Converging plate boundaries are where plates run into each other. The most common type are where oceanic lithosphere subducts. Several mechanisms could contribute to the generation of magmas in this environment.

- Frictional heating is likely to occur along the boundary between the subducted plate and the overlying mantle wedge.
- Flux melting of either the subducted lithosphere or the overlying mantle wedge could occur as a result of the release of volatiles as the subducted plate heats and metamorphoses producing water and/or carbon dioxide fluids.
- The process of subduction may drag the overlying mantle wedge down with it. In order to replace the mantle dragged down in this process, part of the mantle wedge will have to rise. This upwelling of the mantle could result in decompression melting.

If an oceanic lithospheric plate subducts beneath another oceanic lithospheric plate, we find island arcs on the surface above the subduction zone.

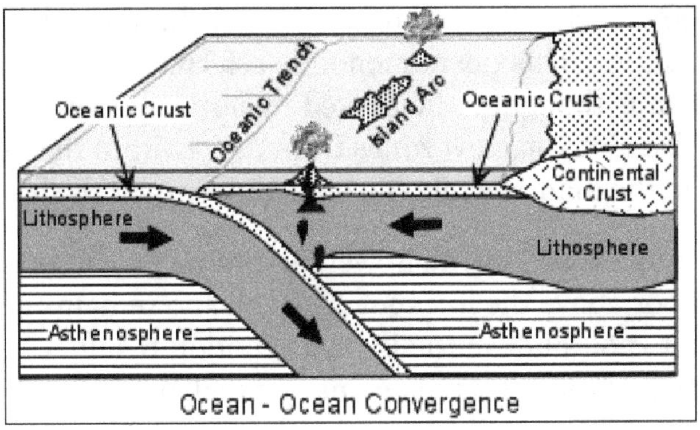

Ocean - Ocean Convergence

If an oceanic plate subducts beneath a plate composed of continental lithosphere, we find continental margin arcs If magma generated near the subduction zone intrudes and cools in the crust, it could induce crustal anatexis.

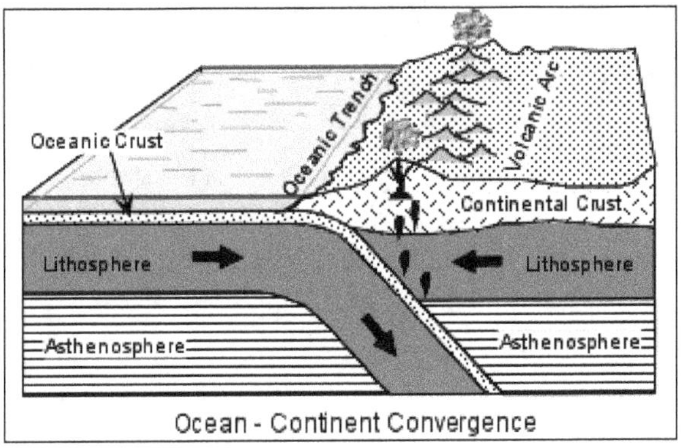

Ocean - Continent Convergence

In areas where two continental lithospheric plates converge fold-thrust mountain ranges develop as the result of compression. If water-bearing crustal rocks are pushed to deeper levels where temperatures are higher, crustal anatexis may result.

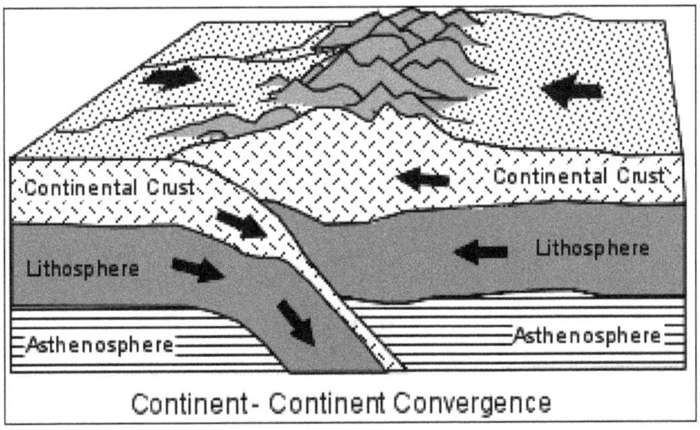

Continent- Continent Convergence

Intraplate Magmatism and Hot Spots

There are a few areas where magmatism does not appear to be related to converging or diverging plate boundaries. These areas occur in the middle of plates, usually far from the plate boundaries. This phenomenon is referred to as intraplate magmatism. Intraplate magmatism is thought to be caused by hot spots formed when thin plumes of mantle material rise along narrow zones from deep within the mantle. The hot spot remains stationary in the mantle while the plate moves over the hot spot.

Decompression melting caused by the upwelling plume produces magmas that form a volcano on the sea floor above the hot spot. The volcano remains active while it is over the vicinity of the hot spot, but eventually plate motion results in the volcano moving away from the plume and the volcano becomes extinct and begins to erode.

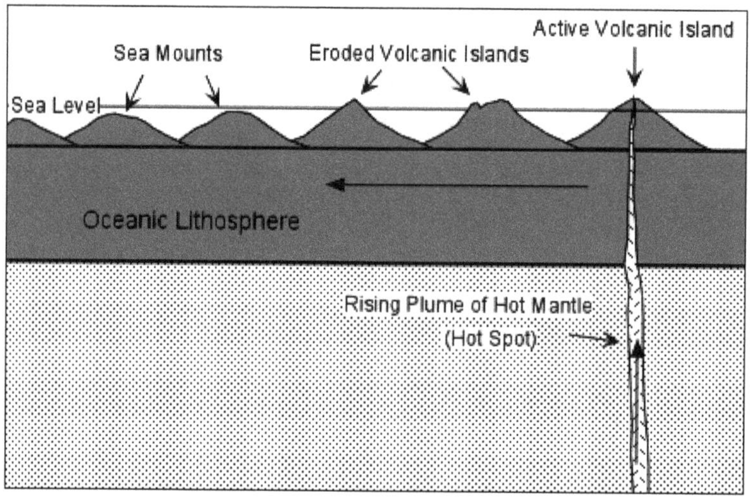

Because the Pacific Plate is one of the faster moving plates, this type of volcanism produces linear chains of islands and seamounts, such as the Hawaiian - Emperor chain, the Line Islands, the Marshall-Ellice Islands, and the Austral seamount chain.

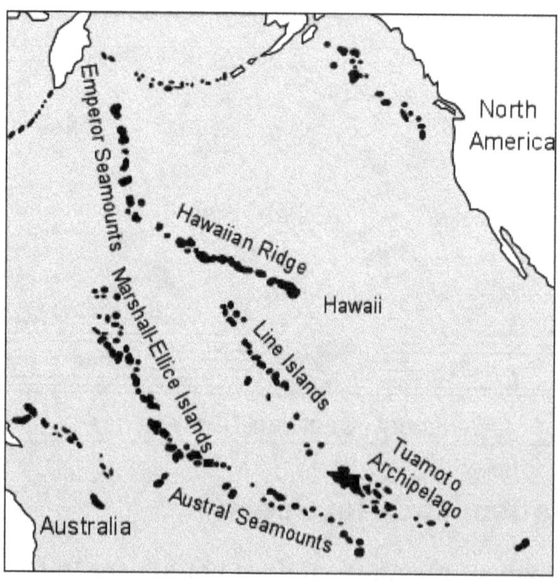

Magmatic Differentiation

Rocks emplaced in any given restricted area during a short amount of geologic time were likely related to the same magmatic event. Evidence for some kind of relationship between the rocks, and therefore between the magmas that cooled to form the rocks came from plotting variation diagrams.

A variation diagram is a plot showing how each oxide component in a rock varies with some other oxide component. Because SiO_2 usually shows the most variation in any given suite of rocks, most variation diagrams plot the other oxides against SiO_2 as shown in the diagram to right, although any other oxide could be chosen for plotting on the x-axis. Plots that show relatively smooth trends of variation of the components suggested that the rocks might be related to one another through some process. Of course, in order for the magmas to be related to one another, they must also have been intruded or erupted within a reasonable range of time. Plotting rocks of Precambrian age along with those of Tertiary age may show smooth variation, but it is unlikely that the magmas were related to one another. If magmas are related to each other by some processes, that process would have to be one that causes magma composition to change.

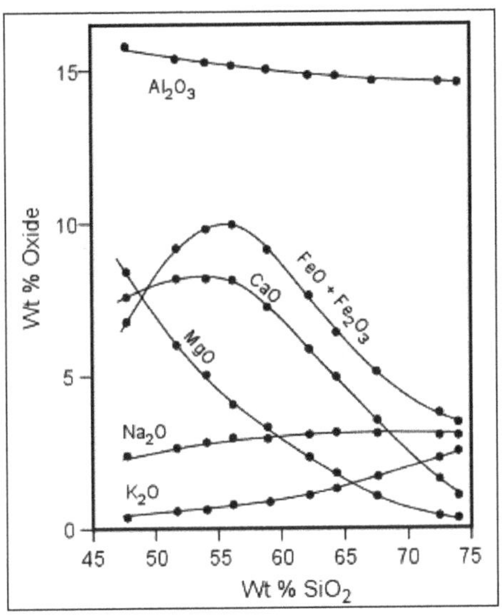

Any process that causes magma composition to change is called magmatic differentiation. Over the years, various process have been suggested to explain the variation of magma compositions observed within small regions. Among the processes are:

- Distinct melting events from distinct sources.
- Various degrees of partial melting from the same source.
- Crystal fractionation.
- Mixing of 2 or more magmas.

- Assimilation/contamination of magmas by crustal rocks.
- Liquid Immiscibility.

Initially, researchers attempted to show that one or the other of these process acted exclusively to cause magmatic differentiation. With historical perspective, we now realize that if any of them are possible, then any or all of these processes could act at the same time to produce chemical change, and thus combinations of these processes are possible.

Distinct Melting Events

One possibility that always exists is that the magmas are not related except by some heating event that caused melting. In such a case each magma might represent melting of a different source rock at different times during the heating event. If this were the case, we might not expect the chemical analyses of the rocks produced to show smooth trends on variation diagrams. But, because variation diagrams based on a closed set of numbers (chemical analyses add up to 100%), if the weight% of one component increases, then the weight percent of some other component must decrease. Thus, even in the event that the magmas are not related SiO_2 could increase and MgO could decrease to produce a trend. The possibility of distinct melting events is not easy to prove or disprove.

Various Degrees of Partial Melting

When a multicomponent rock system melts, unless it has the composition of the eutectic, it melts over a range of temperatures at any given pressure and during this melting the liquid composition changes. Thus, a wide variety of liquid compositions could be made by various degrees of partial melting of the same source rock.

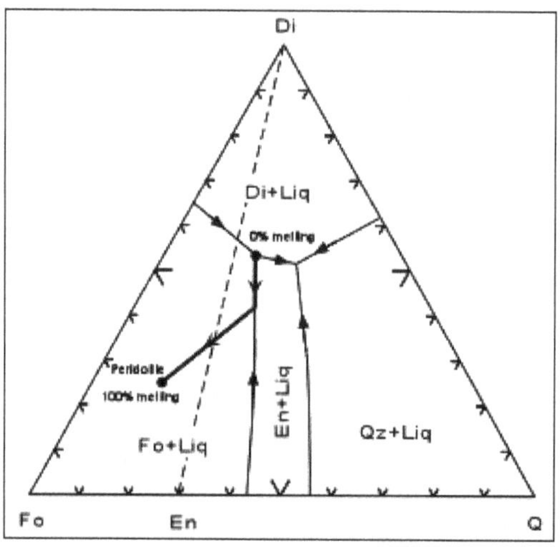

For example: A simple in a three component system containing natural minerals, the system Fo - Di - SiO_2, shown in simplified form here. A proxy for mantle peridotite, being a mixture of Ol, Cpx, and Opx would plot as shown in the diagram. This rock would begin to melt at the peritectic point, where Di, En, Ol, and Liquid are in equilibrium. The composition of the liquid would remain at the peritectic point (labeled 0% melting) until all of the diopside melted. This would occur after about 23% melting. The liquid would then take a path shown by the dark curve, first moving along the En - Ol boundary curve, until the enstatite was completely absorbed, then moving in a direct path toward the peridotite composition.

Fractional Melting

Above some of the liquid must be left behind. If all of the liquid is removed, then we have the case of fractional melting, which is somewhat different.

In fractional melting some all of the liquid is removed at each stage of the process. Let's imagine that we melt the same peridotite again, removing liquids as they form. The first melt to form again will have a composition of the peritectic, labeled "Melt 1" in the diagram. Liquids of composition - Melt 1 can be produced and extracted until all of the Diopside is used up. At this point, there is no liquid, since it has been removed or fractionated, so the remaining solid consists only of Enstatite and Forsterite with composition "Solid 2". This is a two component system. Thus further melting cannot take place until the temperature is raised to the peritectic temperature in the two component system Fo- SiO_2.

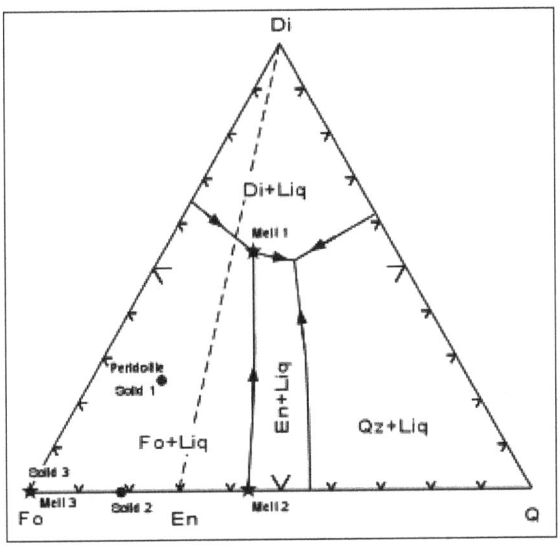

Crystal Fractionation

We saw how liquid compositions can change as a result of removing crystals from the liquid as they form. In all cases, except a eutectic composition, crystallization results in a change in the composition of the liquid, and if the crystals are removed by some process, then different magma compositions can be generated from the initial parent liquid. If minerals that later react to form a new mineral or solid solution minerals are removed, then crystal fractionation can produce liquid compositions that would not otherwise have been attained by normal crystallization of the parent liquid.

Bowen's Reaction Series

Norman L. Bowen, an experimental petrologist in the early 1900s, realized this from his determinations of simple 2- and 3-component phase diagrams, and proposed that if an initial basaltic magma had crystals removed before they could react with the liquid, that the common suite of rocks from basalt to rhyolite could be produced. This is summarized as Bowen's Reaction Series.

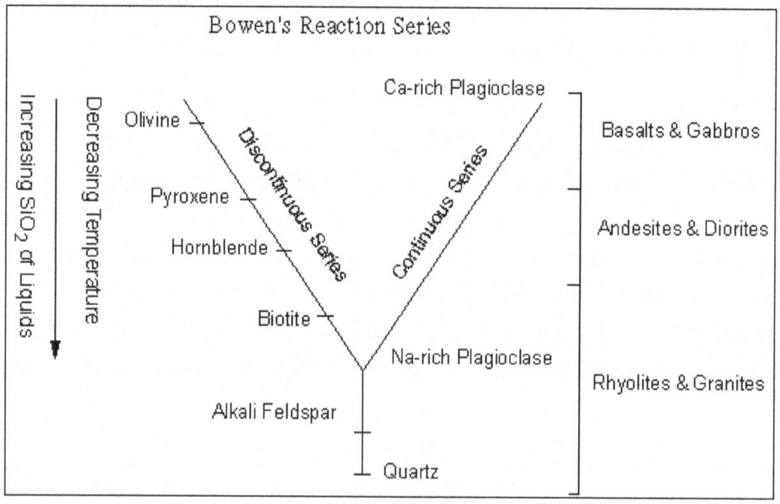

Bowen suggested that the common minerals that crystallize from magmas could be divided into a continuous reaction series and a discontinuous reaction series.

- The continuous reaction series is composed of the plagioclase feldspar solid solution series. A basaltic magma would initially crystallize a Ca- rich plagioclase and upon cooling continually react with the liquid to produce more Na-rich plagioclase. If the early forming plagioclase were removed, then liquid compositions could eventually evolve to those that would crystallize a Na-rich plagioclase, such as a rhyolite liquid.
- The discontinuous reaction series consists of minerals that upon cooling eventually react with the liquid to produce a new phase. Thus, as we have seen, crystallization of olivine from a basaltic liquid would eventually reach a point where olivine would react with the liquid to produce orthopyroxene. Bowen postulated that with further cooling pyroxene would react with the liquid, which by this time had become more enriched in H_2O, to produce hornblende. The hornblende would eventually react with the liquid to produce biotite. If the earlier crystallizing phases are removed before the reaction can take place, then increasingly more siliceous liquids would be produced.

This generalized idea is consistent with the temperatures observed in magmas and with the mineral assemblages we find in the various rocks. We would expect that with increasing SiO_2 oxides like MgO, and CaO should decrease with higher degrees of crystal fractionation because they enter early crystallizing phases, like olivines and pyroxenes. Oxides like H_2O, K_2O and Na_2O should increase with increasing crystal fractionation because they do not enter early crystallizing phases. Furthermore, we would expect incompatible trace element concentrations to increase with fractionation, and compatible trace element concentrations to decrease. This is generally what is observed in igneous rock suites. Because of this, and the fact that crystal fractionation is easy to envision and somewhat easy to test, crystal fraction is often implicitly assumed to be the dominant process of magmatic differentiation.

Mechanisms of Crystal Fractionation

In order for crystal fractionation to operate their must be a natural mechanism that can remove crystals from the magma or at least separate the crystals so that they can no longer react with the liquid. Several mechanisms could operate in nature.

- Crystal Settling/Floating: In general, crystals forming from a magma will have different densities than the liquid.
- If the crystals have a higher density than the liquid, they will tend to sink or settle to the floor of the magma body. The first layer that settles will still be in contact with the magma, but will later become buried by later settling crystals so that they are effectively removed from the liquid.
- If the crystals have a lower density in the magma, they will tend to float or rise upward through the magma. Again the first layer that accumulates at the top of the magma body will initially be in contact with the liquid, but as more crystals float to the top and accumulate, the earlier formed layers will be effectively removed from contact with the liquid.

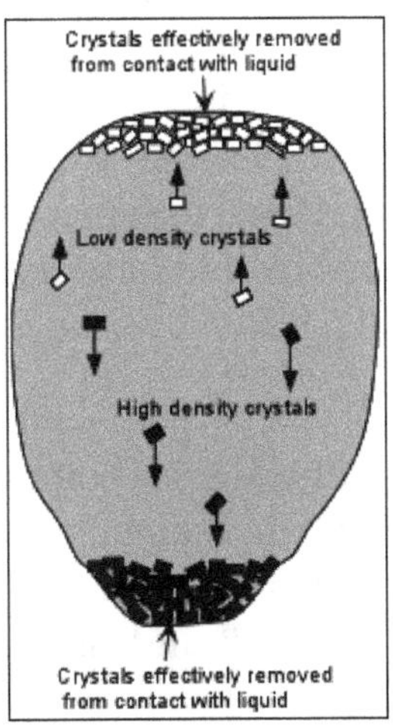

- Inward Crystallization: Because a magma body is hot and the country rock which surrounds it is expected to be much cooler, heat will move outward away from the magma. Thus, the walls of the magma body will be coolest, and crystallization would be expected to take place first in this cooler portion of the magma near the walls. The magma would be expected then to crystallize from the walls inward. Just like in the example above, the first layer of crystals precipitated will still be in contact with the liquid, but will eventually become buried by later crystals and effectively be removed from contact with the liquid.

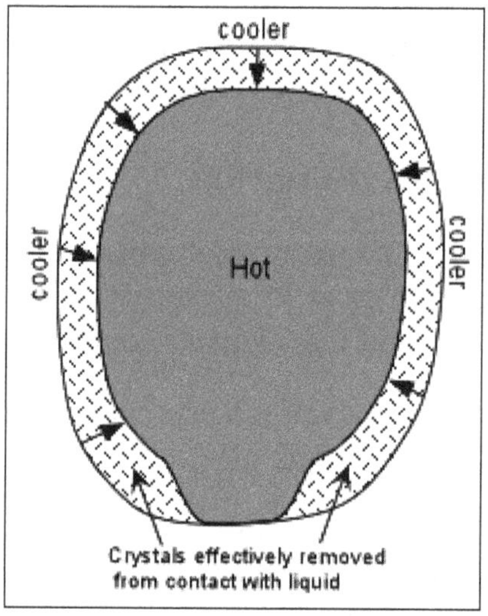

- Filter pressing: This mechanism has been proposed as a way to separate a liquid from a crystal-liquid mush. In such a situation where there is a high concentration of crystals the liquid could be forced out of the spaces between crystals by some kind of tectonic squeezing that moves the liquid into a fracture or other free space, leaving the crystals behind. It would be kind of like squeezing the water out of a sponge. This mechanism is difficult to envision taking place in nature because (1) unlike a sponge the matrix of crystals is brittle and will not deform easily to squeeze the liquid out, and (2) the fractures required for the liquid to move into are generally formed by extensional forces and the mechanism to get the liquid into the fractures involves compressional forces. Filter pressing is a common method used to separate crystals from liquid in industrial processes, but has not been shown to have occurred in nature.

Magma Mixing

If two or more magmas with different chemical compositions come in contact with one another beneath the surface of the Earth, then it is possible that they could mix with to produce compositions intermediate between the end members. If the composition of the magmas is greatly different (i.e. basalt and rhyolite), there are several factors that would tend to inhibit mixing.

- Temperature contrast: Basalts and rhyolites have very different temperatures. If they come in contact with one another the basaltic magma would tend to cool or even crystallize and the rhyolite would tend to heat up and begin to dissolve any crystals that it had precipitated.
- Density Contrast: Basaltic magmas have densities on the order of 2600 to 2700 kg/m^3, whereas rhyolitic magmas have densities of 2300 to 2500 kg/m^3. This contrast in density would mean that the lighter rhyolitic magmas would tend to float on the heavier basaltic magma and inhibit mixing.
- Viscosity Contrast: Basaltic magmas and rhyolitic magmas would have very different viscosities. Thus, some kind of vigorous stirring would be necessary to get the magmas to mix.

Despite these inhibiting factors, there is evidence in rocks that magmas do sometimes mix. The smaller the difference in chemical composition between two magmas, the smaller will be the contrasts in temperature, density, and viscosity.

If magmas of contrasting composition come in contact and begin to mix some kind of stirring mechanism would first be necessary. Such stirring could be provided by convection, Evidence for Mixing:

- Mingling of magmas: If, in the initially stages of such mixing, the magma were erupted, then we might expect to find rocks that show a "marble cake" appearance, with dark colored mafic rock intermingled with lighter colored rhyolitic rock. This, however, is mingling of magmas. Note that differences in color are not always due to differences in composition, so even in rocks that show this banding, mingling of magmas may not have occurred.

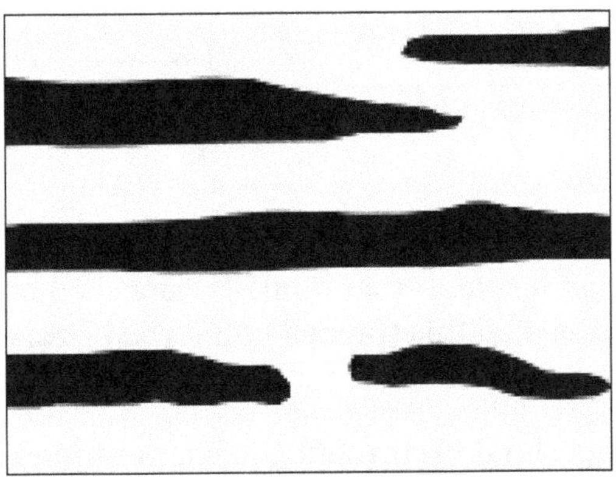

- Disequilibrium Mineral Assemblages: If convective stirring progresses beyond the point of mingling, some evidence might still be preserved if the crystals present in one of the magmas does not completely dissolve or react. This might leave disequilibrium mineral assemblages. For example, if a basaltic magma containing Mg-rich olivine mixed with a rhyolite magma containing quartz and the magma was erupted before the quartz or olivine could be redissolved or made over into another mineral, then we would produce a rock containing mineral that are out of equilibrium.

- Reverse Zoning in Minerals: When a mineral is placed in an environment different than the one in which it originally formed it will tend to react to retain equilibrium. Instead of dissolving completely or remaking their entire composition, solid solution minerals may just start precipitating a new composition that is stable in the new chemical environment or at the new temperature. This can result in zoned crystals that show reversals of the zoning trends. for Example: Mg-Fe solid solution minerals normally zone from Mg-rich cores to Fe-rich rims. If a Fe-rich olivine or pyroxene is mixed into a Mg-rich magma that is precipitating Mg-rich olivine or pyroxene, it may precipitate the more Mg-rich composition on the rims of the added crystals. Analyses of such crystals would reveal a reversal in zoning. Similarly, if a Na-rich plag. originally crystallizing from a rhyolitic magma were mixed into a basaltic magma precipitating a Ca-rich plag., a Ca-rich rim may be added to the Na-rich plagioclase.

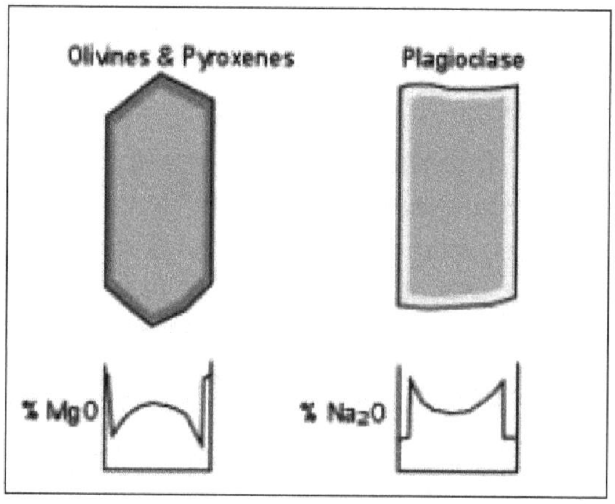

Glass Inclusions

Crystal growth from liquids is sometimes not perfect. Sometimes the crystal grows incompletely trapping liquid inside. If that liquid is quenched on the surface and a thin section is cut through the crystal this trapped liquid will be revealed as glass inclusions in the crystal.

Since the glass inclusions should represent the composition of the magma that precipitated the crystal, chemical analysis of glass inclusions give us the composition of the liquid in which the crystal formed. The groundmass may also contain glass representing the composition of the liquid in which the crystal resided at eruption. If the composition of glass inclusions is different from glass in the groundmass, and if the groundmass composition is not what is expected from normal crystallization of the minerals present, this provides evidence of magma mixing.

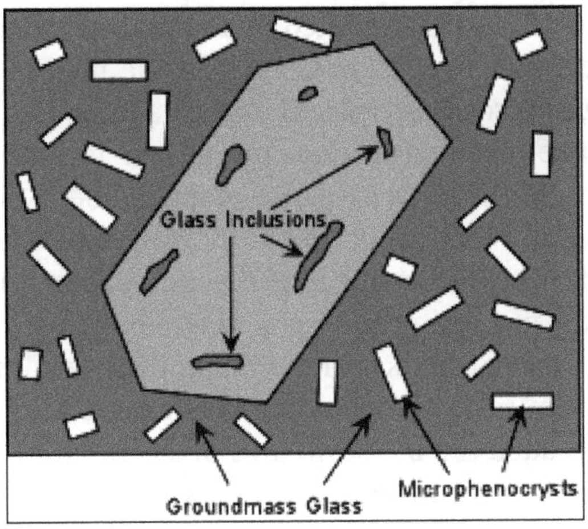

Chemical Evidence

If the mixing process proceeds to the point where other evidence is erased, evidence for mixing will still be preserved in the composition of the mixed magmas. On oxide-oxide variation diagrams mixtures will lie along a straight line. Thus if diagrams show a group of rocks that lie along the same straight line, and the proportional distances are the same on all diagrams, one could hypothesize that the chemical variation resulted from magma mixing.

Crustal Assimilation/Contamination

Because the composition of the crust is generally different from the composition of magmas which must pass through the crust to reach the surface, their is always the possibility that reactions between the crust and the magma could take place. If crustal rocks are picked up, incorporated into the magma, and dissolved to become part of the magma, we say that the crustal rocks have been assimilated by the magma. If the magma absorbs part of the rock through which it passes we say that the magma has become contaminated by the crust. Either of these process would produce a change in the chemical composition of the magma unless the material being added has the same chemical composition as the magma.

In a sense, bulk assimilation would produce some of the same effects as mixing, but it is more complicated than mixing because of the heat balance involved. In order to assimilate the country rock enough heat must be provided to first raise the country rock to its solidus temperature where it will begin to melt and then further heat must be added to change from the solid state to the liquid state. The only source of this heat, of course, is the magma itself.

Evidence for Assimilation/Contamination

As magma passes upward through the crust pieces of the country rock through which it passes may be broken off and assimilated by the magma. Just as in magma mixing, various stages of this process may be preserved in the magma and rock that results. Xenoliths (meaning foreign rock) are pieces of rock sometimes found as inclusions in other rocks. The presence of xenoliths does not always indicate that assimilation has taken place, but if the xenoliths show evidence of having been disaggregated with their minerals distributed thought the rest of the rock it is likely that some contamination of the magma has taken place. This may result in disequilibrium mineral assemblages and reversely zoned minerals, just as in the case of magma mixing. And, if the assimilation goes to completion, with all of the xenoliths being dissolved in the magma, the only evidence left may be chemical, and again similar to the straight line mixing patterns produced by mixing. Perhaps the best evidence of assimilation/contamination comes from studies of radiogenic isotopes. Here we give an example using the systematics of the Rb - Sr system.

- ^{87}Rb is a radioactive isotope that decays to ^{87}Sr with a half life of billion years.
- Because Rb is an incompatible element, it has been extracted from the mantle by magmas and added to the crust. Thus the concentration of Rb in the crust (avg. about 100 ppm) is much higher than it is in the mantle (avg. about 4 ppm).
- 86Sr is a stable, non radiogenic isotope whose concentration does not change with time.
- Because ^{87}Rb decays to produce ^{87}Sr and because there is more Rb in the crust than in the mantle, the ^{87}Sr/ ^{87}Sr of the crust has, over time changed to much higher values than the ^{87}Sr/^{87}Sr ratio in the mantle.
- The ^{87}Sr/^{87}Sr ratio of the mantle is generally in the range between 0.702 - 0.705. Thus rocks derived from melting of the mantle should have ^{87}Sr/^{87}Sr ratios in this range.
- ^{87}Sr/^{87}Sr ratios of crustal rocks will depend on their age and concentration of Rb. Older crustal rocks will have high values of ^{87}Sr/^{87}Sr in the range 0.705 - 0.720, younger crustal rocks having been recently derived from the mantle will ^{87}Sr/^{87}Sr ratios more similar to the mantle.
- If mantle derived magmas assimilate or are contaminated by older crustal rocks, then we would expect to find ratios of ^{87}Sr/^{87}Sr in these contaminated rocks that are higher than those found in the mantle and extend up to values found in older crustal rocks.

Liquid Immiscibility

Liquid immiscibility is where liquids do not mix with each other. We're all familiar with this phenomenon in the case of oil and water/vinegar in salad dressing. For example: In the alkali feldspar system. Just like in the alkali feldspar system, immiscibility is temperature dependent.

For example: In a two component system if there is a field of immiscibility it would appear as in the diagram to shown here. Cooling of a liquid with a composition of 25%B & 75%A would eventually result in the liquid separating into two different compositions. Further with further cooling these liquids one liquid would become more enriched in A and the other more enriched in B. Eventually both liquids would reach a temperature where crystals of A would start to form. Both liquids would be in equilibrium with crystals of A at the same temperature. Further cooling would result in the disappearance of the A-rich liquid.

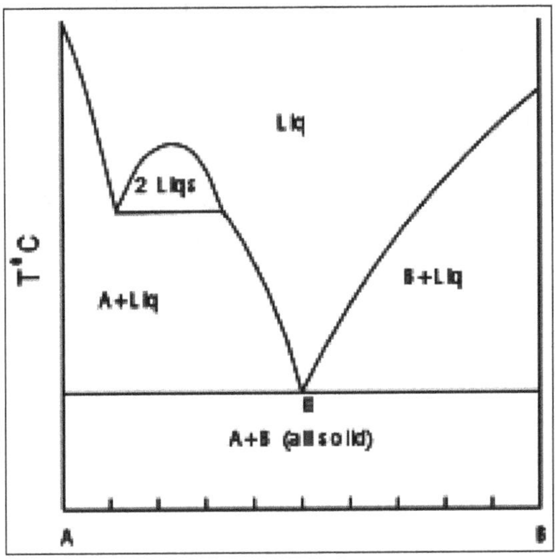

This point out two important properties of immiscible liquids:

- If immiscible liquids are in equilibrium with solids, both liquids must be in equilibrium with the same solid compositions.
- Extreme compositions of the two the liquids will exist at the same temperature.

Liquid immiscibility was once thought to be a mechanism to explain all magmatic differentiation. If so, requirement 2, would require that siliceous liquids and mafic liquids should form at the same temperature. Since basaltic magmas are generally much hotter than rhyolitic magmas, liquid immiscibility is not looked upon favorably as an explanation for wide diversity of magmatic compositions. Still, liquid immiscibility is observed in experiments conducted on simple rock systems.

Combined Processes

If any of these process are possible, then a combination of the process could act to produce chemical change in magmas. Thus, although crystal fractionation seems to be the dominant process affecting magmatic differentiation, it may not be the only processes. As we have seen, assimilation is likely to accompanied by crystallization of magmas in order to provide the heat necessary for assimilation. If this occurs then a combination of crystal fraction and assimilation could occur. Similarly, magmas could mix and crystallize at the same time resulting in a combination of magma mixing and crystal fractionation. In nature, things could be quite complicated.

Forms of Igneous Rocks

The Igneous rock masses which were produced due to the consolidation of the magma within the country rocks and solidifying underneath the surface of the earth are known as intrusive bodies. Magma moves through joints, fractures and between the crystals of the solid rock of the crust and mantle. When it reaches its solidification temperature, it crystallises. Dykes, sills, and plutons are igneous bodies that have cooled from magma beneath the surface. If the magma crystallises at depths of 20 to 30km it is called a plutonic rock and will have large crystals. If the magma crystallises at a shallow depth below the surface it will have small to medium crystals. The deeper the emplacement of the magma the slower will be the crystallisation and bigger will be the crystals formed, similarly the shallower the depth of emplacement, the faster the magma cools and crystallise forming smaller crystals.

The intrusive rocks are characterised by the development of a variety of forms, depending upon their exact mode of emplacement and relation with the surrounding country-rocks. The magma will follow a route which is at the least pressure, usually along fractures or cracks.

Chilled Margin: When magma cool and solidified, they will crystallise fastest where they are in contact with the colder country rock. Crystals in the centre will be larger than those occurring on the edge of the intrusion. The outside edge with the smallest crystals is called the chilled margin. Form depends on mode of emplacement within the country-rocks.

In Unfolded Regions

- Discordant: If an intrusive mass happens to cut across the structures of the preexisting rocks of the country it is said to be a discordant intrusion. The preexisting rocks can be igneous, sedimentary or metamorphic.
- Concordant: In a similar manner, the concordant rock masses necessarily run parallel to the structure of the country rocks in which they occur. If the country rock be made up of sedimentary bedding, a concordant intrusive mass should run parallel to the bedding planes while a discordant body must cut across them. If metamorphic rocks are forming the country, concordant bodies must

run parallel to the foliation or lineation structures of the country rock, on the other hand, the discordant masses should cut them through these structures. In the case of igneous country rocks, all other subsequent bodies of igneous origin are likely to be discordant in nature since the igneous country-rock doesn't ordinarily exhibit any characteristic features to which the later injections may run parallel.

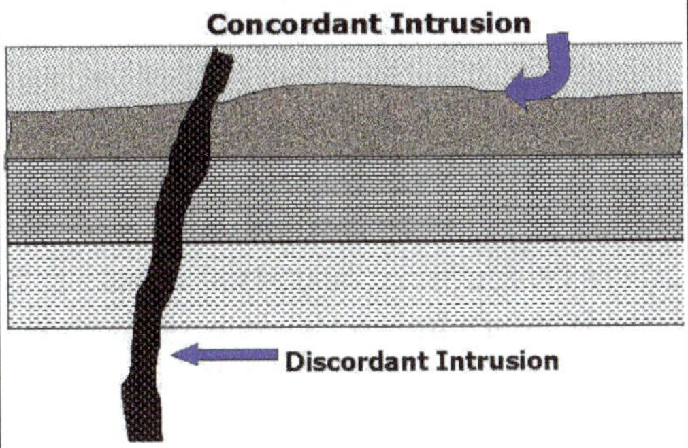

Showing concordant and discordant intrusions.

Showing discordant structures (dyke and lava flows).

- Dykes: Dykes are discordant igneous bodies of more or less tabular shape and they have cross cutting relationship with the country-rocks. Dykes are commonly formed due to the development of cracks within the country rocks and their subsequent filling up with the igneous intrusives. Dykes commonly occur in the coal fields and associated tracts in India, e.g. Lamprophyric rocks broadly represented by kimberlites.

Showing discordant intrusions (e.g. dolerite dyke cuts across the gabbroic rock bodies).

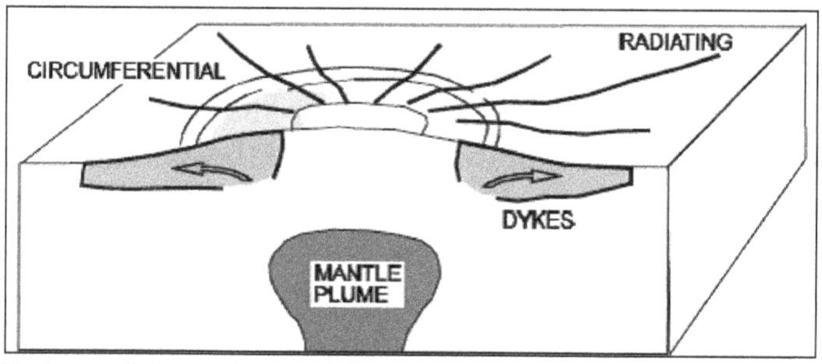

Giant radiating and circumferential dyke swarms generated above a mantle plume.

Dykes can be occurring in groups, may exhibit radiating arcuate and any other pattern. An arcuate dyke occurring more or less in the form of a complete or nearly complete circle may be called a ring-dyke. Dyke of arcuate outcrop occurring more or less in the form of a complete circle. Sometimes sheets of igneous materials are found to occur in the form of inverted, coaxial cones, with thin layers of country-rocks, that sheets may be designated as cone sheets. These structures are formed due to rise and fall of pressure of magma within the subcrustal magma chambers with consequent development of cracks along suitable directions.

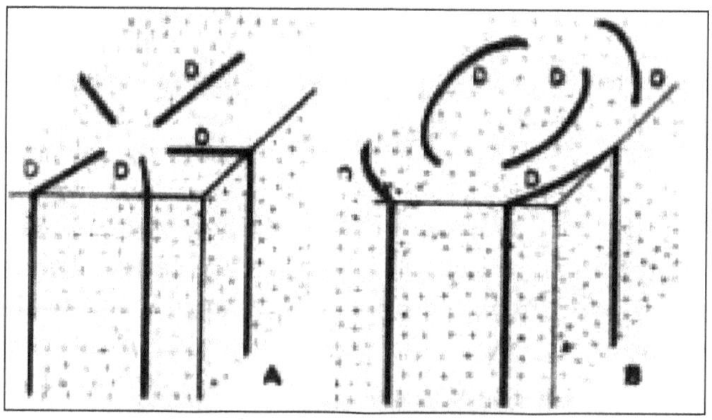

Radiating and Arcuate dykes intrusions.

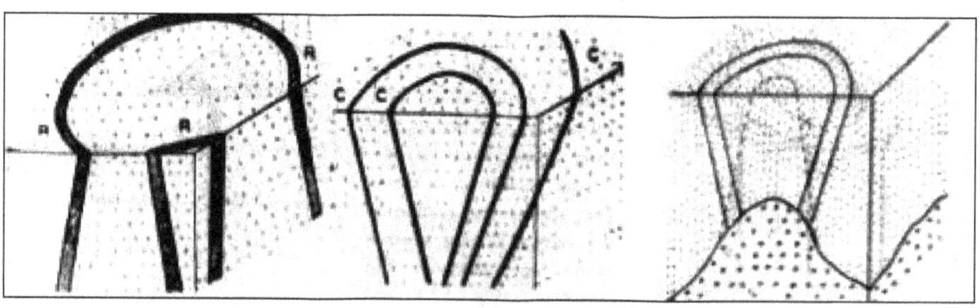
Ring Dyke and Cone Sheet.

- Sills: Sills are tabular concordant igneous rock bodies lying parallel to the structure of the country rocks. They are commonly sheet like masses, occurring within the country rocks, having a variable thickness. The mode of occurrence is parallel to the bedding planes of the sedimenrty rock series itself, in which they actually occur as intrusive bodies. They are commonly horizontal or nearly so their attitude but they can have other orientation in space depending on the attitude and nature of the rock beds in which they are associated. Lamprophyres and mica-peridotite sills occur in the coalfields of the Damodar area. Dykes and sills commonly occur in mutual association, a sill originating from dykes and dykes from sills and it appears as a network of intrusions within the country rocks at shallow depth.

Showing concordant igneous rocks masses in terms of sills.

In the Highly Folded Regions

- Laccoliths: Laccoliths are concordant igneous bodies with their lower surface flat and upper surface arched in the form of dome such bodies are naturally formed due to the accumulation of viscous magma underneath the rocks occurring upon the surface. Laccoliths may have dykes or sill acting as their feeders.
- Lapoliths: Lapoliths are concordant igneous bodies with their shape that of a central part of the mass is thicker while it thins out towards the periphery it is produced due to the collapse of the rock beds in beds in any region and consequent up-rise of the magma through the collapsing layers.

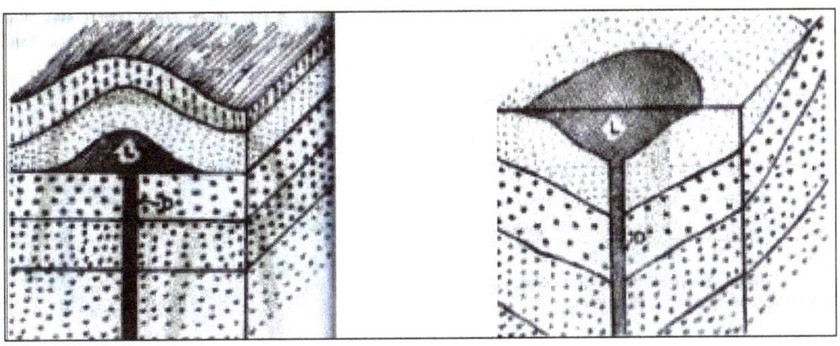

The body marked with 'L' is a Laccolith (first figure) and Lapolith in the second figure and 'D' are the dykes which act as its feeder.

- Batholiths: Batholiths are very large intrusive bodies, usually so large and it is very rare to exposed at the bottom. Sometimes they are composed of several smaller intrusions. The actual mode of emplacement of batholiths can follow the two schools of thought. According to one school, the batholiths are supposed to have been formed due to the forceful intrusion of the magma within the country rocks. Another school, however, is of the opinion that the emplacement was a comparatively slow process. Field evidence has suggested that the existence of both the above kinds of modes of the emplacement. Batholiths are formed, necessarily under deep-seated, plutonic conditions, commonly during periods of mountain-building and are found exposed on the surface of the earth only after prolonged and conspicuous erosion, the best example being Ladakh and Karakoram batholiths.

- Stocks: Stocks are smaller bodies of intrusion that are likely to feed from the deeper level of the batholiths. Stocks may have been feeders for volcanic eruptions, but because large amounts required exposing a stock or batholith, the associated volcanic rocks are rarely exposed.

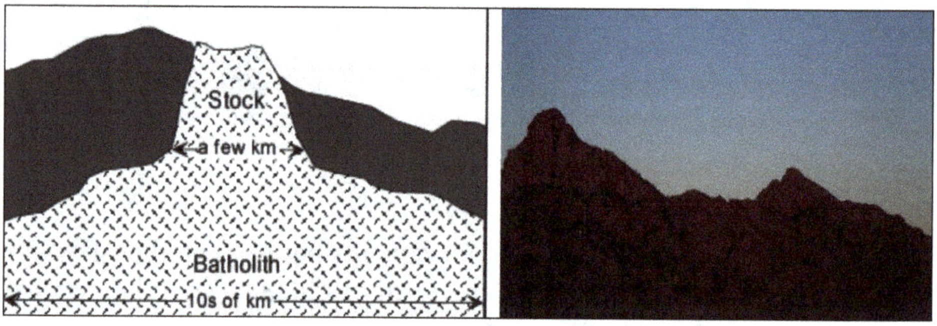

Batholith and stock igneous intrusions.

- Phacoliths: Phacoliths are the concordant plutonic bodies that lie parallel to the bedding plane or foliation of folded country rocks. They occur along the crests of anticlines or troughs of the syncline in folded sedimentary strata.

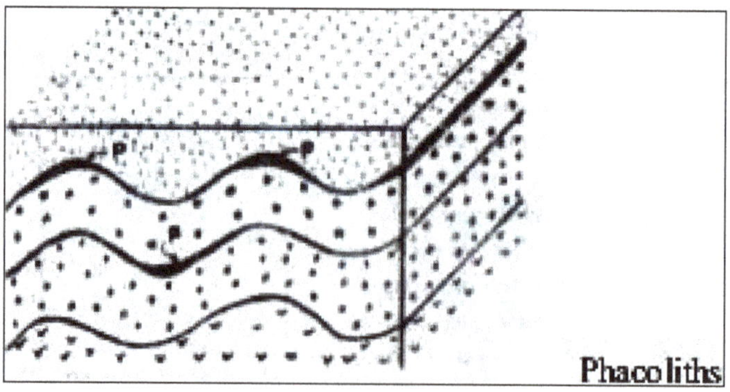

Plutons

A pluton is a rock body composed of intrusive igneous rocks that have been formed or crystallized when magma slowly cooled below the surface of the earth. The term pluton is commonly used for larger igneous bodies, especially when the shape and structural relation are not clear. Plutons are of all sizes and shapes from those a few feet long to those that are exposed over hundreds of square miles. Thus it includes dikes, sills, laccoliths, lopoliths, batholiths, and other igneous bodies.

For the structural study of plutons following factors are involved:

- Internal Structure,
- Size and Shape,
- Chronological and structural relationships to the adjacent rocks.

Internal Structure

The rocks in plutons may range from fine grained to very coarse grained. It depends upon its cooling history and volatiles present during consolidation and crystallizing power of various mineral species. Many igneous rocks are massive showing no preferred orientation of mineral constituents. Some are characterized by foliation resulting from the parallel arrangement of platy or ellipsoidal mineral grains. Banded or layered intrusive rocks are those that consist of alternate layer of different mineral composition.

Figure shows (a) course grained rock characterized by foliation resulting from the parallel arrangement of platy or ellipsoidal mineral grains (b) Massive rocks which showing no preferred orientation of mineral constituents. (c &d) Course grained a banded and layered intrusive rock that consists of an alternate layer of different mineral composition.

- Inclusions: We may recognize the fragments of older rock (angular sub-angular or round) surrounded by igneous rock. A xenolith is an inclusion that has been derived from country rocks or some older formation. Such as sandstone found in granite, which is commonly genetically unrelated to igneous rock. An autolith is an inclusion of an older igneous rock, genetically related to younger igneous rock in which it occurs. An inclusion of diorite in granodiorite is autolith.

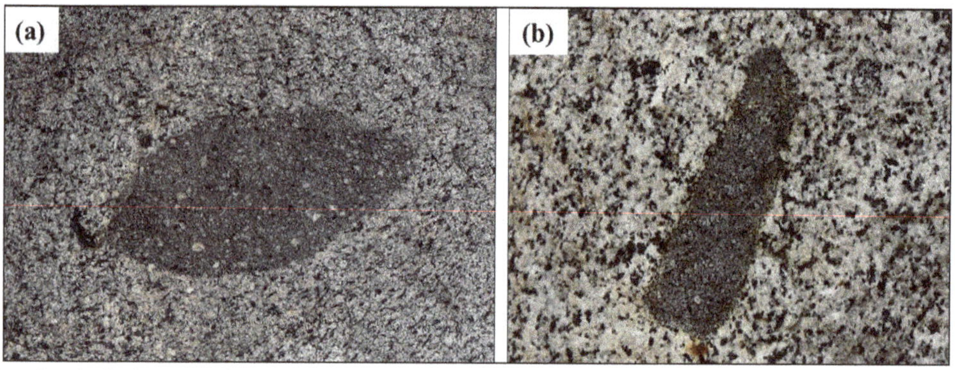

(a) Showing inclusions sandstone found in granite (b) An inclusion of diorite in granodiorite is autolith.

- Segregation: A rounded or irregular body with few inches to many feet in diameter and in some cases hundreds of feet across, that has been enriched in one or more of the mineral composing the igneous rocks. e.g. the hornblende granite may contain clots that have much hornblende than the surrounding granite.
- Schlieren: Wavy streaky, irregular sheets which are lacking sharp contacts with surrounding igneous rocks, based on compositions it can be darker or lighter in which they occur.

Showing Schlieren inclusion.

Shape and Size of Plutons

Larger plutons it is quite difficult to examine the complete shape directly, but the smaller shape and size of plutons may be observed directly. Many dikes and sills are smaller than in which they occur. Although, complete shape of the entire original body can never be determined because the part above of the surface is forever lost due to erosion and below the surface it is determined only by a very expensive programme of drilling and excavation.

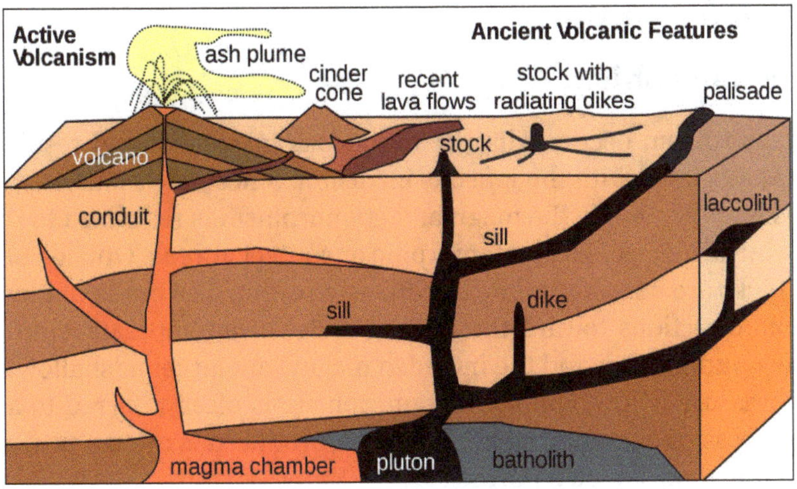

showing all types of igneous inclusions.

Metamorphic Petrology

The mineralogical and structural adjustment of solid rocks to physical and chemical conditions that have been imposed at depths below the near surface zones of weathering and diagenesis and which differ from conditions under which the rocks in question originated.

Metamorphic rocks can be derived from sedimentary, igneous or other metamorphic rocks from heat and pressure in which the rock remains in a solid state (melting would create and igneous rock). In general terms metamorphic rocks are characterised by an increase in both crystal size and density though there are some exceptions.

Grade of Metamorphism

As the temperature and/or pressure increases on a body of rock we say the rock undergoes prograde metamorphism or that the grade of metamorphism increases. Metamorphic grade is a general term for describing the relative temperature and pressure conditions under which metamorphic rocks form.

- Low-grade metamorphism takes place at temperatures between about 200 to 320 °C, and relatively low pressure. Low grade metamorphic rocks are generally characterized by an abundance of hydrous minerals. With incasing grade of metamorphism, the hydrous minerals begin to react with other minerals and/or break down to less hydrous minerals.
- High-grade metamorphism takes place at temperatures greater than 320 °C and relatively high pressure. As grade of metamorphism increases, hydrous minerals become less hydrous, by losing H_2O, and non-hydrous minerals become more common.

Types of Metamorphism

Contact Metamorphism

Contact metamorphism occurs adjacent to igneous intrusions and results from high temperatures associated with the igneous intrusion. Since only a small area surrounding the intrusion is heated by the magma, metamorphism is restricted to the zone surrounding the intrusion, called a metamorphic or contact aureole Outside of the contact aureole, the rocks are not affected by the intrusive event. The grade of metamorphism increases in all directions toward the intrusion. Because temperature differences between the surrounding rock and the intruded magma are larger at shallow levels in the crust where pressure is low, contact metamorphism is often referred to as high temperature, low pressure metamorphism. The rock produced is often a fine-grained rock that shows no foliation, called a hornfels.

Regional Metamorphism

Regional metamorphism occurs over large areas and generally does not show any relationship to igneous bodies. Most regional metamorphism is accompanied by deformation under non-hydrostatic or differential stress conditions. Thus, regional metamorphism usually results in forming metamorphic rocks that are strongly foliated, such as slates, schists, and gniesses. The differential stress usually results from tectonic forces that produce compressional stresses in the rocks, such as when two continental masses collide. Thus, regionally metamorphosed rocks occur in the cores of fold/thrust mountain belts or in eroded mountain ranges. Compressive stresses result in folding of rock and thickening of the crust, which tends to push rocks to deeper levels where they are subjected to higher temperatures and pressures.

Cataclastic Metamorphism

Cataclastic metamorphism occurs as a result of mechanical deformation, like when two bodies of rock slide past one another along a fault zone. Heat is generated by the friction of sliding along such a shear zone, and the rocks tend to be mechanically deformed, being crushed and pulverized, due to the shearing. Cataclastic metamorphism is not very common and is restricted to a narrow zone along which the shearing occurred.

Hydrothermal Metamorphism

Rocks that are altered at high temperatures and moderate pressures by hydrothermal fluids are hydrothermally metamorphosed. This is common in basaltic rocks that generally lack hydrous minerals. The hydrothermal metamorphism results in alteration to such Mg-Fe rich hydrous minerals as talc, chlorite, serpentine, actinolite, tremolite, zeolites, and clay minerals.

Burial Metamorphism

When sedimentary rocks are buried to depths of several hundred meters, temperatures greater than 300 °C may develop in the absence of differential stress. New minerals grow, but the rock does not appear to be metamorphosed. The main minerals produced are the Zeolites. Burial metamorphism overlaps, to some extent, with diagenesis, and grades into regional metamorphism as temperature and pressure increase.

Shock Metamorphism

When an extraterrestrial body, such as a meteorite or comet impacts with the Earth or if there is a very large volcanic explosion, ultrahigh pressures can be generated in the impacted rock. These ultrahigh pressures can produce minerals that are only stable at very high pressure, such as the SiO_2 polymorphs coesite and stishovite. In addition they can produce textures known as shock lamellae in mineral grains, and such textures as shatter cones in the impacted rock.

Classification of Metamorphic Rocks

Rocks can be classified into Mineral assemblage, texture, protolith, and bulk chemical composition of the rock.

Texture

In metamorphic rocks individual minerals may or may not be bounded by crystal faces. Those that are bounded by their own crystal faces are termed idioblastic. Those that show none of their own crystal faces are termed xenoblastic. From examination of metamorphic rocks, it has been found that metamorphic minerals can be listed in a generalized sequence, known as the crystalloblastic series, listing minerals in order of their tendency to be idioblastic. In the series, each mineral tends to develop idioblastic surfaces against any mineral that occurs lower in the series. This series is listed below:

- Rutile, Sphene, Magnetite.
- Tourmaline Kyanite, Staurolite, Garnet, Andalusite.
- Epidote, Zoisite, Lawsonite, Forsterite.
- Pyroxenes, Amphiboles, Wollastonite.
- Micas, Chlorites, Talc, Stilpnomelane, Prehnite.
- Dolomite, Calcite.
- Scapolite, Cordierite, Feldspars.
- Quartz.

This series can, in a rather general way, enable us to determine the origin of a given rock. For example a rock that shows euhedral plagioclase crystals in contact with anhedral amphibole, likely had an igneous protolith, since a metamorphic rock with the same minerals would be expected to show euhedral amphibole in contact with anhedral plagioclase.

Another aspect of the crystalloblastic series is that minerals high on the list tend to form porphyroblasts (the metamorphic equivalent of phenocrysts), although K-feldspar (a minerals that occur lower in the list) may also form porphyroblasts. Porphyroblasts are often riddled with inclusions of other minerals that were enveloped during growth of the porphyroblast. These are said to have a poikioblastic texture.

Most metamorphic textures involve foliation. Foliation is generally caused by a preferred orientation of sheet silicates. If a rock has a slatey cleavage as its foliation, it is termed a slate, if it has a phyllitic foliation, it is termed a phyllite, if it has a shistose foliation, it is termed a schist. A rock that shows a banded texture without a distinct foliation is termed a gneiss. All of these could be porphyroblastic (i.e. could contain porhyroblasts). A rock that shows no foliation is called a hornfels if the grain size is small, and a granulite, if the grain size is large and individual minerals can be easily distinguished with a hand lens.

Protolith

Protolith refers to the original rock, prior to metamorphism. In low grade metamorphic rocks, original textures are often preserved allowing one to determine the likely protolith. As the grade of metamorphism increases, original textures are replaced with metamorphic textures and other clues, such as bulk chemical composition of the rock, are used to determine the protolith.

Bulk Chemical Composition

The mineral assemblage that develops in a metamorphic rock is dependent on:

- The pressure and temperature reached during metamorphism.
- The composition of any fluid phase present during metamorphism.
- The bulk chemical composition of the rock.

Just like in igneous rocks, minerals can only form if the necessary chemical constituents are present in the rock (i.e. the concept of silica saturation and alumina saturation applies to metamorphic rocks as well). Based on the mineral assemblage present in the rock one can often estimate the approximate bulk chemical composition of the rock. Some terms that describe this general bulk chemical composition are as follows:

- Pelitic: These rocks are derivatives of aluminous sedimentary rocks like shales and mudrocks. Because of their high concentrations of alumina they are recognized by an abundance of aluminous minerals, like clay minerals, micas, kyanite, sillimanite, andalusite, and garnet.
- Quartzo-Feldspathic: Rocks that originally contained mostly quartz and feldspar like granitic rocks and arkosic sandstones will also contain an abundance of quartz and feldspar as metamorphic rocks, since these minerals are stable of a wide range of temperature and pressure. Those that exhibit mostly quartz and feldspar with only minor amounts of aluminous minerals are termed quartzo-feldspathic.
- Calcareous: Calcareous rocks are calcium rich. They are usually derivatives of carbonate rocks, although they contain other minerals that result from reaction of the carbonates with associated siliceous detrital minerals that were present in the rock. At low grades of metamorphism calcareous rocks are recognized by their abundance of carbonate minerals like calcite and dolomite. With increasing grade of metamorphism these are replaced by minerals like brucite, phlogopite (Mg-rich biotite), chlorite, and tremolite. At even higher grades anhydrous minerals like diopside, forsterite, wollastonite, grossularite, and calcic plagioclase.

- Basic: Just like in igneous rocks, the general term basic refers to low silica content. Basic metamorphic rocks are generally derivatives of basic igneous rocks like basalts and gabbros. They have an abundance of Fe-Mg minerals like biotite, chlorite, and hornblende, as well as calcic minerals like plagioclase and epidote.
- Magnesian: Rocks that are rich in Mg with relatively less Fe, are termed magnesian. Such rocks would contain Mg-rich minerals like serpentine, brucite, talc, dolomite, and tremolite. In general, such rocks usually have an ultrabasic protolith, like peridotite, dunite, or pyroxenite.
- Ferriginous: Rocks that are rich in Fe with little Mg are termed ferriginous. Such rocks could be derivatives of Fe-rich cherts or ironstones. They are characterized by an abundance of Fe-rich minerals like greenalite (Fe-rich serpentine), minnesotaite (Fe-rich talc), ferroactinolite, ferrocummingtonite, hematite, and magnetite at low grades, and ferrosilite, fayalite, ferrohedenbergite, and almandine garnet at higher grades.
- Manganiferrous: Rocks that are characterized by the presence of Mn-rich minerals are termed manganiferrous. They are characterized by such minerals as Stilpnomelane and spessartine.

What is Visible in the Rock and its Degree of Metamorphism

Classification is generally loose and practical such that names can be adapted to describe the rock in the most satisfactory way that conveys the important characteristics. Three kinds of criteria are normally employed. These are:

- Mineralogical: The most distinguishing minerals are used as a prefix to a textural term. Thus, a schist containing biotite, garnet, quartz, and feldspar, would be called a biotite-garnet schist. A gneiss containing hornblende, pyroxene, quartz, and feldspar would be called a hornblende-pyroxene gneiss. A schist containing porphyroblasts of K-feldspar would be called a K-spar porphyroblastic schist.
- Chemical: If the general chemical composition can be determined from the mineral assemblage, then a chemical name can be employed. For example a schist with a lot of quartz and feldspar and some garnet and muscovite would be called a garnet-muscovite quartzo-feldspathic schist. A schist consisting mostly of talc would be called a talc-magnesian schist.
- Protolithic: If a rock has undergone only slight metamorphism such that its original texture can still be observed then the rock is given a name based on its original name, with the prefix meta-applied. For example: metabasalt, metagraywacke, meta-andesite, metagranite.

There are certain non-foliated rocks with specific chemical compositions and/or mineral assemblages. These rocks are given specific names:

- Amphibolites: These are medium to coarse grained, dark colored rocks whose principal minerals are hornblende and plagioclase. They result from metamorphism of basic igneous rocks. Foliation is highly variable, but when present the term schist can be appended to the name (i.e. amphibolite schist).
- Marbles: These are rocks composed mostly of calcite, and less commonly of dolomite. They result from metamorphism of limestones and dolostones. Some foliation may be present if the marble contains micas.
- Eclogites: These are medium to coarse grained consisting mostly of garnet and green clinopyroxene called omphacite, that result from high grade metamorphism of basic igneous rocks. Eclogites usually do not show foliation.
- Quartzites: Quartz arenites and chert both are composed mostly of SiO_2. Since quartz is stable over a wide range of pressures and temperatures, metamorphism of quartz arenites and cherts will result only in the recrystallization of quartz forming a hard rock with interlocking crystals of quartz. Such a rock is called a quartzite.
- Serpentinites: Serpentinites are rocks that consist mostly of serpentine. These form by hydrothermal metamorphism of ultrabasic igneous rocks.
- Soapstones: Soapstones are rocks that contain an abundance of talc, which gives the rock a greasy feel, similar to that of soap. Talc is an Mg-rich mineral, and thus soapstones from ultrabasic igneous protoliths, like peridotites, dunites, and pyroxenites, usually by hydrothermal alteration.
- Skarns: Skarns are rocks that originate from contact metamorphism of "dirty" limestones or dolostones, and show evidence of having exchanged constituents with the intruding magma. Thus, skarns are generally composed of minerals like calcite and dolomite, from the original carbonate rock, but contain abundant calcium and magnesium silicate minerals like andradite, grossularite, epidote, vesuvianite, diopside, and wollastonite that form by reaction of the original carbonate minerals with silica from the magma. The chemical exchange is that takes place is called metasomatism.
- Mylonites: Mylonites are cataclastic metamorphic rocks that are produced along shear zones deep in the crust. They are usually fine-grained, sometimes glassy, that are streaky or layered, with the layers and streaks having been drawn out by ductile shear.

Metamorphic Facies

In general, metamorphic rocks do not drastically change chemical composition during metamorphism, except in the special case where metasomatism is involved (such as in the production of skarns,). The changes in mineral assemblages are due to changes in the temperature and pressure conditions of metamorphism. Thus, the mineral assemblages that are observed must be an indication of the temperature and pressure environment that the rock was subjected to. This pressure and temperature environment is referred to as Metamorphic Facies. (This is similar to the concept of sedimentary facies, in that a sedimentary facies is also a set of environmental conditions present during deposition.). The sequence of metamorphic facies observed in any metamorphic terrain, depends on the geothermal gradient that was present during metamorphism.

- A high geothermal gradient might be present around an igneous intrusion, and would result in metamorphic rocks belonging to the hornfels facies.
- Under a normal to high geothermal gradient rocks would progress from zeolite facies to greenschist, amphibolite, and eclogite facies as the grade of metamorphism (or depth of burial) increased.
- If a low geothermal gradient was present then rocks would progress from zeolite facies to blueschist facies to eclogite facies.

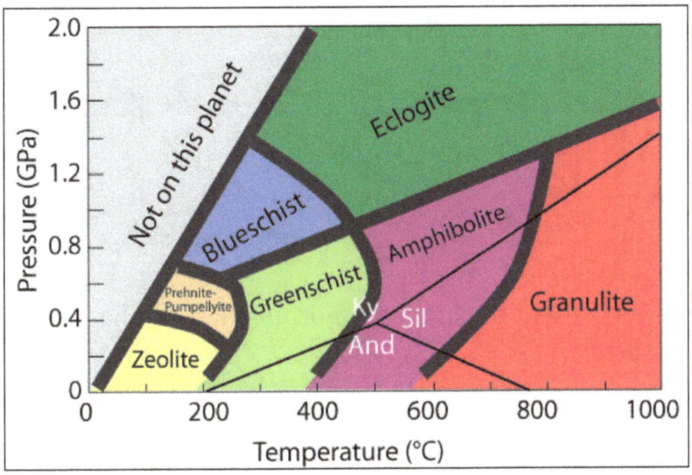

Thus, if we know the facies of metamorphic rocks in the region, we can determine what the geothermal gradient must have been like at the time the metamorphism occurred.

Metamorphic Textures

Metamorphic rocks exhibit a variety of textures. These can range from textures similar to the original protolith at low grades of metamorphism, to textures that are purely produced during metamorphism and leave the rock with little resemblance to the original protolith.

Foliation is defined as a pervasive planar structure that results from the nearly parallel alignment of sheet silicate minerals and/or compositional and mineralogical layering in the rock. Most foliation is caused by the preferred orientation of phylosilicates, like clay minerals, micas, and chlorite. Preferred orientation develops as a result of non-hydrostatic or differential stress acting on the rock (also called deviatoric stress).

Pressure increases with depth of burial, thus, both pressure and temperature will vary with depth in the Earth. Pressure is defined as a force acting equally from all directions. It is a type of stress, called hydrostatic stress or uniform stress. If the stress is not equal from all directions, then the stress is called a differential stress. Normally geologists talk about stress as compressional stress. Thus if a differential stress is acting on the rock, the direction along which the maximum principal stress acts is called s_1, the minimum principal stress is called s_3, and the intermediate principal stress direction is called s_2. Note that extensional stress would act along the direction of minimum principal stress.

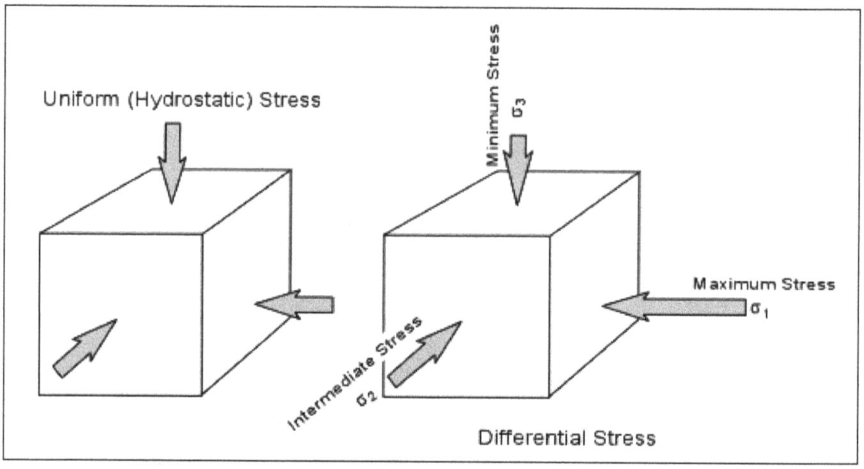

If differential stress is present during metamorphism, it can have a profound effect on the texture of the rock.

Rounded grains can become flattened in the direction of maximum compressional stress. Minerals that crystallize or grow in the differential stress field may develop a preferred orientation. Sheet silicates and minerals that have an elongated habit will grow with their sheets or direction of elongation orientated perpendicular to the direction of maximum stress. This is because growth of such minerals is easier along directions parallel to sheets, or along the direction of elongation and thus will grow along s_3 or s_2, perpendicular to s_1.

Since most phyllosilicates are aluminous minerals, aluminous (pelitic) rocks like shales, generally develop a foliation as the result of metamorphism in a differential stress field. Example: Metamorphism of a shale (made up initially of clay minerals and quartz) Shales have fissility that is caused by the preferred orientation of clay minerals with their {001} planes orientated parallel to bedding. Metamorphic petrologists and structural geologists refer to the original bedding surface S_0.

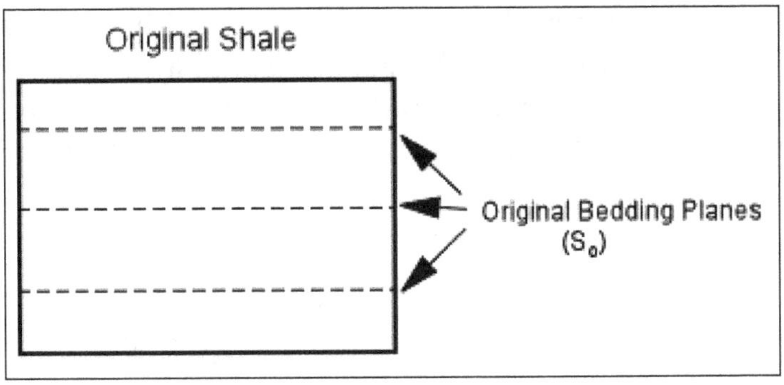

Slate

Slates form at low metamorphic grade by the growth of fine grained chlorite and clay minerals. The preferred orientation of these sheet silicates causes the rock to easily break planes parallel to the sheet silicates, causing a slatey cleavage. Note that in the case shown here, the maximum principle stress is oriented at an angle to the original bedding planes so that the slatey cleavage develops at an angle to the original bedding. The foliation or surface produced by this deformation is referred to S_1.

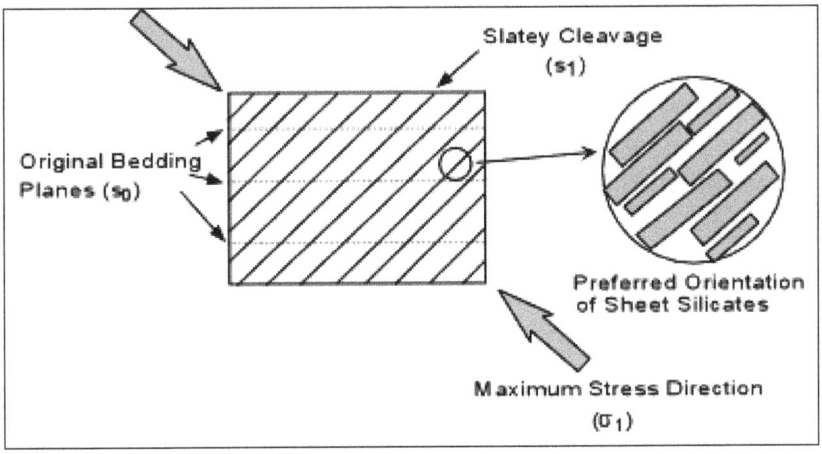

Schist

The size of the mineral grains tends to enlarge with increasing grade of metamorphism. Eventually the rock develops a near planar foliation caused by the preferred orientation of sheet silicates (mainly biotite and muscovite). Quartz and feldspar grains, however show no preferred orientation. The irregular planar foliation at this stage is called schistosity.

Gneiss

As metamorphic grade increases, the sheet silicates become unstable and dark colored minerals like hornblende and pyroxene start to grow. These dark colored minerals tend to become segregated into distinct bands through the rock (this process is called metamorphic differentiation), giving the rock a gneissic banding. Because the dark colored minerals tend to form elongated crystals, rather than sheet- like crystals, they still have a preferred orientation with their long directions perpendicular to the maximum differential stress.

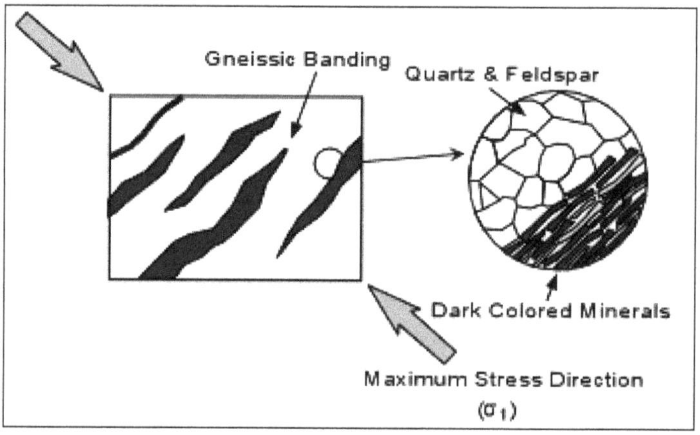

Granulite

At the highest grades of metamorphism most of the hydrous minerals and sheet silicates become unstable and thus there are few minerals present that would show a preferred orientation. The resulting rock will have a granulitic texture that is similar to a phaneritic texture in igneous rocks.

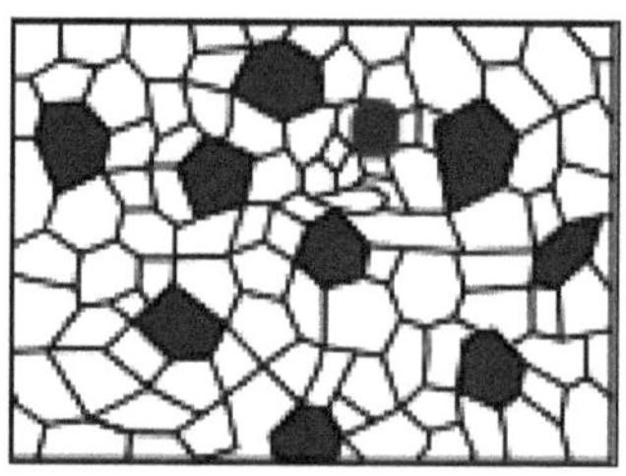

In general, the grain size of metamorphic rocks tends to increase with increasing grade of metamorphism, as seen in the progression form fine grained shales to coarser (but still fine) grained slates, to coarser grained schists and gneisses.

Metamorphism and Deformation

Most regionally metamorphosed rocks (at least those that eventually get exposed at the Earth's surface) are metamorphosed during deformational events. Since deformation involves the application of differential stress, the textures that develop in metamorphic rocks reflect the mode of deformation, and foliations or slatey cleavage that develop during metamorphism reflect the deformational mode and are part of the deformational structures.

The deformation involved in the formation of fold-thrust mountain belts generally involves compressional stresses. The result of compressional stress acting on rocks that behave in a ductile manner (ductile behavior is favored by higher temperature, higher confining stress [pressure] and low strain rates) is the folding of rocks. Original bedding is folded into a series of anticlines and synclines with fold axes perpendicular to the direction of maximum compressional stress. These folds can vary in their scale from centimeters to several kilometers between hinges. Note that since the axial planes are oriented perpendicular to the maximum compressional stress direction that slatey cleavage or foliation should also develop along these directions. Thus, slatey cleavage or foliation is often seen to be parallel to the axial planes of folds, and is sometimes referred to axial plane cleavage or foliation.

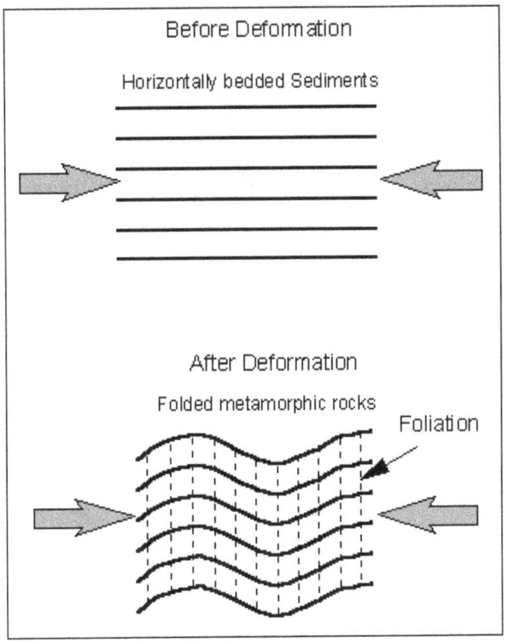

Metamorphic Differentiation

Gneisses, and to some extent schists, show compositional banding or layering, usually evident as alternating somewhat discontinuous bands or layers of dark colored ferromagnesian minerals and lighter colored quartzo-feldspathic layers. The development of such compositional layering or banding is referred to as metamorphic differentiation. Throughout the history of metamorphic petrology, several mechanisms have been proposed to explain metamorphic differentiation.

Preservation of Original Compositional Layering

In some rocks the compositional layering may not represent metamorphic differentiation at all, but instead could simply be the result of original bedding. For example, during the early stages of metamorphism and deformation of interbedded sandstones and shales the compositional layering could be preserved even if the maximum compressional stress direction were at angle to the original bedding.

In such a case, a foliation might develop in the shale layers due to the recrystallization of clay minerals or the crystallization of other sheet silicates with a preferred orientation controlled by the maximum stress direction.

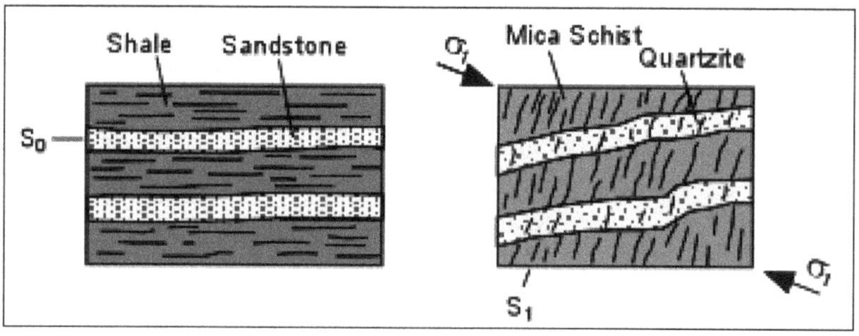

In such a case, it would be easy to determine that the compositional layers represented original bedding because the foliation would cut across the compositional layering.

In highly deformed rocks that have undergone both folding and shearing, it may be more difficult to determine that the compositional layering represents original bedding. As shearing stretches the bedding, individual folded beds may be stretched out and broken to that the original folds are not easily seen.

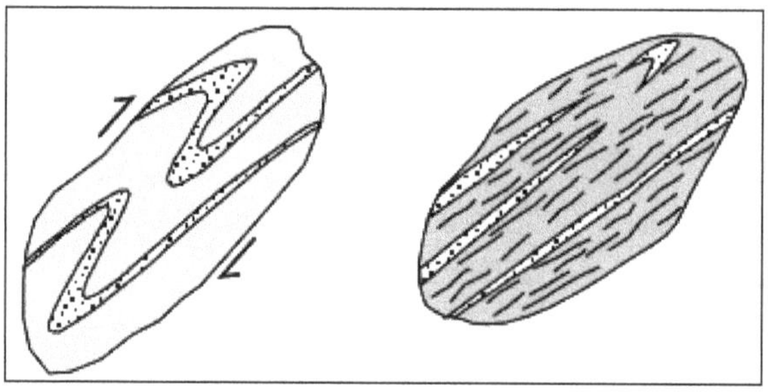

Similarly, if the rock had been injected by dikes or sills prior to metamorphism, these contrasting compositional bands, not necessarily parallel to the original bedding, could be preserved in the metamorphic rock.

Transposition of Original Bedding

Original compositional layering a rock could also become transposed to a new orientation during metamorphism. The diagram below shows how this could occur. In the initial stages a new foliation begins to develop in the rock as a result of compressional stress at some angle to the original bedding. As the minerals that form this foliation grow, they begin to break up the original beds into small pods. As the pods are compressed and extended, partly by recrystallization, they could eventually intersect again to form new compositional bands parallel to the new foliation.

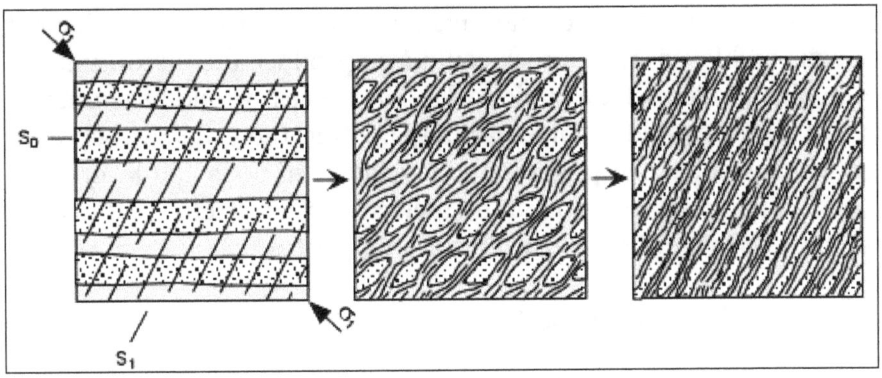

Solution and Reprecipitation

In fine grained metamorphic rocks small scale folds, called kink bands, often develop in the rock as the result of application of compressional stress. A new foliation begins to develop along the axial planes of the folds. Quartz and feldspar may dissolve as a result of pressure solution and be reprecipitated at the hinges of the folds where the pressure is lower. As the new foliation begins to align itself perpendicular to s_1, the end result would be alternating bands of micas or sheet silicates and quartz or feldspar, with layering parallel to the new foliation.

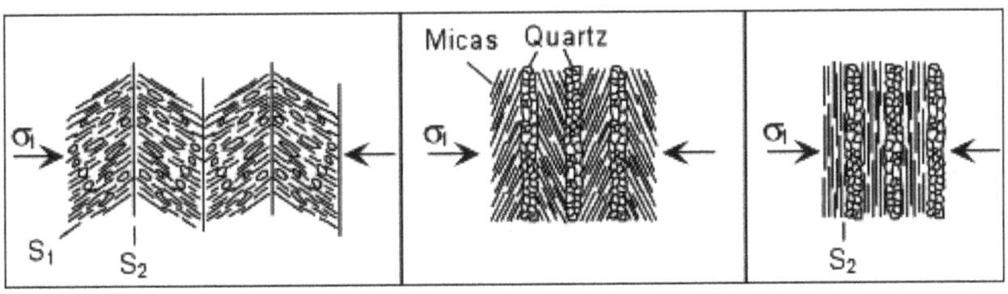

Preferential Nucleation

Fluids present during metamorphism have the ability to dissolve minerals and transport ions from one place in the rock to another. Thus felsic minerals could be dissolved from one part of the rock and preferentially nucleate and grow in another part of the rock to produce discontinuous layers of alternating mafic and felsic compositions.

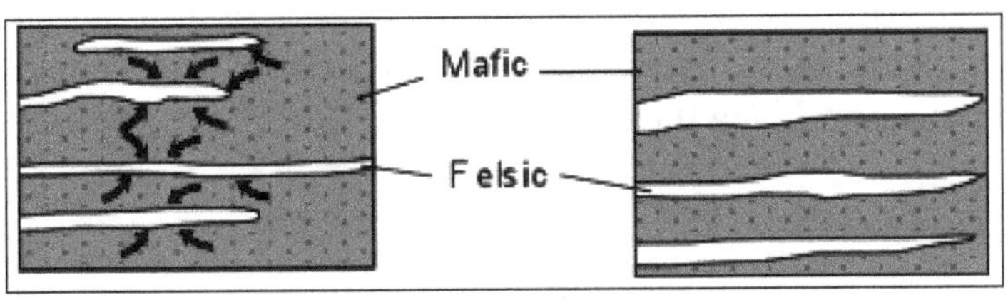

Migmatisation

Migmatites are small pods and lenses that occur in high grade metamorphic terranes that may represent melts of the surrounding metamorphic rocks. Injection of the these melts into pods and layers in the rock could also produce the discontinuous banding often seen in high grade metamorphic rocks. The process would be similar to that described except that it would involve partially melting the original rock to produce a felsic melt, which would then migrate and crystallise in pods and layers in the metamorphic rock. Further deformation of the rock could then stretch and fold such layers so that they may no longer by recognizable as migmatites.

Metamorphic Minerals/Reactions

The mineral assemblages that occur in metamorphic rocks depend on four factors:

- The bulk chemical composition of the original rock.
- The pressure reached during metamorphism.
- The temperature reached during metamorphism.
- The composition of any fluid phase that was present during metamorphism.

If a rock is taken to some higher pressure and temperature then the mineral assemblage that develops should represent stable chemical equilibrium if the conditions are held for a long enough period of time that equilibrium can be achieved. Since metamorphism usually involves long periods of geologic time, most metamorphic rocks represent an equilibrium mineral assemblage. If a rock is taken to some higher pressure and temperature then the mineral assemblage that develops should represent stable chemical equilibrium if the conditions are held for a long enough period of time that equilibrium can be achieved. Since metamorphism usually involves long periods of geologic time, most metamorphic rocks represent an equilibrium mineral assemblage.

Progressive/Prograde Metamorphism

Progressive or Prograde metamorphism occurs as the temperature and pressure are increased on the rock.As the pressure and temperature increase, a rock of a given chemical composition is expected to undergo a continuous series of chemical reactions between its constituent minerals and any fluid phase present to produce a series of new mineral assemblages that are stable at the higher pressures and temperatures. Here is how the mineral assemblages might change in a hypothetical set of rocks, starting with a low grade mineral assemblage then medium grade then high grade.

Low Grade Rocks

First, consider the compositions of various types of rocks. Pelitic and quartzo-feldspathic rocks are relatively aluminous, calcareous rocks are relatively rich in the C component, and basic rocks are relative rich in the F component. At some low temperature and pressure typical of low grade metamorphism, pelitic and quartzo feldspathic rocks in this example would consist of epidote, chlorite, andalusite, muscovite, and possibly quartz, k-spar, muscovite, and albite.Calcareous rocks would contain epidote, calcite, actinolite, and possibly quartz, albite, and k-spar. Basic rocks would consist of epidote, chlorite, actinolite, and possibly, quartz, albite, and k-spar. Ultramafic rocks would consist of actinolite, chlorite and talc.

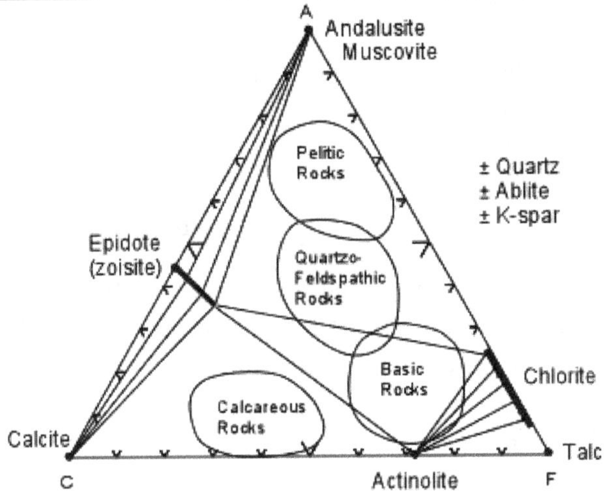

Low Grade Rocks

Medium Grade Rocks

At some higher pressure and temperature the assemblage might change as follows: pelitic rocks would consist of sillimanite, plagioclase, cordierite, muscovite, and possibly quartz and k-spar (the albite is now included in the plagioclase). Quartzo-feldspathic rocks would consist of plagioclase, cordierite, biotite, and likely quartz and k-spar. Calcareous rocks would consist of grossularite, calcite and hornblende. Basic rocks would contain plagioclase, biotite, hornblende, and perhaps quartz and k-spar. Ultramafic rocks would consist of biotite anthopyllite, and hornblende.

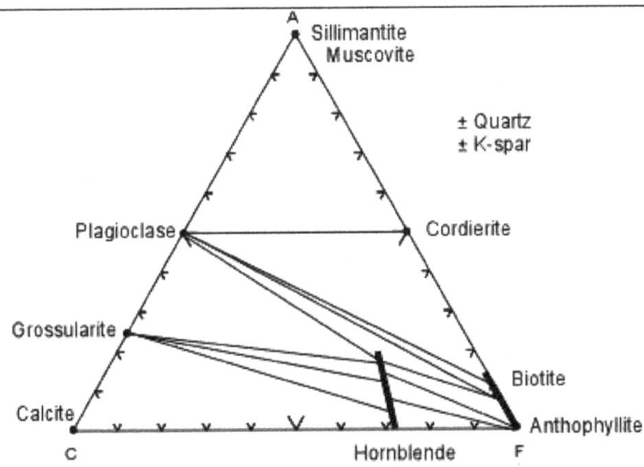

Higher Grade Rocks

Now let's consider what happened in jumping from the pressure/temperature conditions of the first diagram to those of the second diagram. First, we list the phases that have disappeared and those that have appeared.

Phases disappeared	New phases
Andalusite	Sillimanite
Epidote/zoisite	Plagioclase
Chlorite	Grossularite
Talc	Cordierite
Actinolite	Biotite
Albite	Anthophyllite
	Hornblende

Here are the chemical reactions that must have occurred between the two sets of pressure/temperature conditions that explain the new mineral assemblages. For the disappearance of andalusite and appearance of sillimanite the reaction is simple:

$$\underset{\text{Andalusite}}{Al_2SiO_5} \Rightarrow \underset{\text{Sillimanite}}{Al_2SiO_5}$$

Epidote (zoisite) would break down to produce the anorthite component of plagioclase and grossularite, resulting in the expulsion of water as a fluid phase:

$$4\underset{\text{Epidote/Zoisite}}{Ca_2Al_3Si_3O_{12}(OH)} + \underset{\text{Qtz}}{SiO_2} \Rightarrow 5\underset{\text{Anorthite (plag)}}{CaAl_2Si_2O_8} + \underset{\text{grossularite}}{Ca_3Al_2Si_3O_{12}} + 4\underset{\text{fluid}}{H_2O}$$

Chlorite could have reacted with muscovite and quartz to produce biotite, cordierite, and fluid:

$$\underset{\text{Chlorite}}{(Mg,Fe)_5Al_2Si_3O_{10}(OH)_2} + \underset{\text{Muscovite}}{KAl_3Si_3O_{10}(OH)_2} + 2\underset{\text{Qtz}}{SiO_2} \Rightarrow$$

$$\underset{\text{Biotite}}{K(Fe,Mg)_3AlSi_3O_{10}(OH)_2} + \underset{\text{Cordierite}}{(Mg,Fe)_2Al_4Si_5O_{18}} + 4\underset{\text{fluid}}{H_2O}$$

Talc would break down to produce anthophyllite, quartz, and fluid:

$$7\underset{\text{Talc}}{Mg_3Si_4O_{10}(OH)_8} \Rightarrow 3\underset{\text{Anthophyllite}}{Mg_7Si_8O_{22}(OH)_2} + 4\underset{\text{Qtz}}{SiO_2} + 4\underset{\text{fluid}}{H_2O}$$

and chlorite, actinolite, epidote (zoisite), and quartz would react to produce hornblende and fluid:

$$7\underset{\text{Chlorite}}{(Mg,Fe)_5Al_2Si_3O_{10}(OH)_2} + 13\underset{\text{Actinolite}}{Ca_2(Mg,Fe)_5Si_8O_{22}(OH)_2}$$

$$+ 12\underset{\text{epidote (Zoisite)}}{Ca_2Al_3Si_3O_{12}(OH)} + 14\underset{\text{Qtz}}{SiO_2} \Rightarrow$$

$$\underset{\text{Hornblende}}{25Ca_2(Mg,Fe)_4Al_2Si_7O_{22}(OH)_2} + \underset{\text{fluid}}{22H_2O}$$

With the exception of the first reaction (andalusite ⇒ sillimanite) all of these reactions result in the evolution of H_2O in a fluid phase. Such reactions are called dehydration reactions.

High Grade Rocks

Increase the temperature and pressure so that a new set of minerals develops for each rock. At these new pressure/temperature conditions, pelitic rocks would contain plagioclase, sillimanite, cordierite, quartz, and k-spar. Quartzo-feldspathic rocks would consist of plagioclase, cordierite, almandine/pyrope garnet, quartz, and k-spar. Calcareous rocks would consist either of wollastonite, grossularite and diopside, or grossularite, plagioclase, and diopside, or plagioclase, diopside, and biotite, depending on the composition of the initial rock. Basic rocks would consist of plagioclase, diopside, and almandine/pyrope, or plagioclase, hypersthene, and almandine/pyrope. Ultramafic rocks would consist of hypersthene, almandine/pyrope, and diopside.

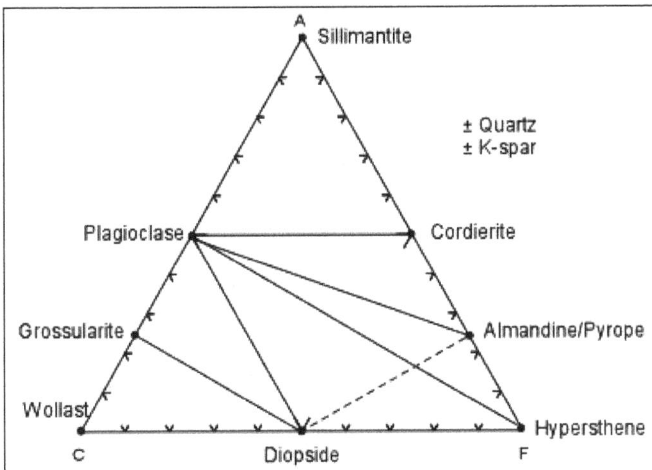

Highest Grade Rocks

A list of the phases that disappeared and those that appeared at some point between the two pressure/temperature conditions include.

Phases disappeared	New phases
Muscovite	Wollastonite
Calcite	Diopside
Hornblende	Hypersthene
Anthophyllite	Almandine/pyrope
Biotite	Hypersthene and k-spar

Here are the chemical reactions that must have occurred between the two sets of pressure/temperature conditions that explain the new mineral assemblages. For the disappearance of muscovite:

$$\underset{\text{Muscovite}}{KAl_3Si_3O_{10}(OH)_2} + \underset{\text{Qtz}}{SiO_2} \Rightarrow \underset{\text{K-spar}}{KAlSi_3O_8} + \underset{\text{Sillimanite}}{Al_2SiO_5} + \underset{\text{fluid}}{H_2O}$$

For the disappearance of calcite and the appearance of wollastonite:

$$\underset{\text{Calcite}}{CaCO_3} + \underset{\text{Qtz}}{SiO_2} \Rightarrow \underset{\text{Wollastonite}}{CaSiO_3} + \underset{\text{fluid}}{CO_2}$$

Hornblende break down to diopside, hypersthene, plagioclase, and fluid:

$$\underset{\text{Hornblende}}{Ca_2Mg_4Al_2Si_7O_{22}(OH)_2} \Rightarrow \underset{\text{Diopside}}{CaMgSi_2O_6} + \underset{\text{Hypersthene}}{3MgSiO_3}$$

$$+ \underset{\text{Plagioclase}}{CaAl_2Si_2O_8} + \underset{\text{fluid}}{H_2O}$$

Anthophyllite would break down to hypersthene and quartz:

$$\underset{\text{Anthophyllite}}{Mg_7Si_8O_{22}(OH)_2} \Rightarrow \underset{\text{Hypersthene}}{7MgSiO_3} + \underset{\text{Qtz}}{SiO_2} + \underset{\text{fluid}}{H_2O}$$

For the breakdown of biotite to form hypersthene and k-spar the following reaction must have occurred:

$$\underset{\text{Biotite}}{K(Fe,Mg)_3\,AlSi_3O_{10}(OH)_2} + \underset{\text{Qtz}}{3SiO_2} \Rightarrow \underset{\text{K-spar}}{KAlSi_3O_8} + \underset{\text{Hypersthene}}{3(Mg,Fe)SiO_3} + \underset{\text{fluid}}{H_2O}$$

For the formation of almandine/pyrope from cordierite and anthophyllite:

$$\underset{\text{Cordierite}}{(Mg,Fe)_2\,Al_4Si_5O_{18}} + \underset{\text{Anthophyllite}}{(Mg,Fe)_7\,Si_8O_{22}(OH)_2} \Rightarrow$$

$$\underset{\text{Almandine/Pyrope}}{2(Mg,Fe)\;Al_2Si_3O_{12}} + \underset{\text{Hypersthene}}{3(Mg,Fe)SiO_3} + \underset{\text{Qtz}}{4SiO_2} + \underset{\text{fluid}}{H_2O}$$

All of the reactions evolve a fluid phase. The calcite $\Rightarrow$ wollastonite reaction is a decarbonation reaction, and the rest are dehydration reactions. Prograde metamorphism involves a series of devolatilization reactions that produce a fluid phase. The fluid phase generally will consist of a mixture of H_2O and CO_2. This results in producing minerals that are less volatile rich with increasing grade of metamorphism. Ultimately, at the highest grade of metamorphism no hydrous or carbonate bearing phases exist.

Retrograde Metamorphism

If retrograde metamorphism were a common process then upon uplift and unroofing metamorphic rocks would progressively return to mineral assemblages stable at lower pressures and temperatures. Yet, high grade metamorphic rocks are common at the surface of the Earth and usually show only minor retrograde minerals.

Three factors inhibit retrograde metamorphism, two of which involve the fluid phase.

- Chemical reactions run faster at higher temperatures: During prograde metamorphism the reaction boundaries are continually being over-stepped to higher temperature. At higher temperature diffusion rates are higher and molecular vibration required to break chemical bonds is higher. Thus, during prograde metamorphism reaction rates are faster. As temperature is lowered on a rock, the reaction boundaries are over-stepped to low temperature, and as a result the reaction rates are much slower. Of course, this depends on the rate at which temperature is lowered. It is certainly possible for temperatures to decrease at such a slow rate during uplift and unroofing, that retrograde mineral assemblages would have time to form. Also, a second episode of prograde metamorphism may occur during which temperatures are increased again. Minerals stable at the new temperature may start to grow and overprint the minerals that were formed at higher temperature during the first episode of metamorphism.
- A fluid phase is driven off as a result of the devolatilization reactions: during prograde metamorphism, As pressure increases, porosity of rocks also decreases, and thus this fluid phase will likely be driven out of the rock body. In the absence of the fluid phase it is impossible to form hydrous minerals and carbonates, since H_2O and CO_2, two of the key components needed in such reactions, may not be present.
- The fluid phase also helps to catalyze chemical reactions: Although the net reactions may appear to be solid-solid reactions, in reality there may be more involved. For example the fluid phase could dissolve a mineral in one part of the rock and precipitate a new mineral in another part of the rock, just as happens during diagenesis of sedimentary rocks. If the fluid phase is driven off during prograde metamorphism, then it will not be available to catalyze the reactions to produce the retrograde mineral assemblage as pressure and temperature are lowered.

Concept and Classification of Metamorphic Facies and Facies Series

Agents of Metamorphism

The extraneous factors that ultimately bring about metamorphic changes in crustal rocks are 'Pressure', 'Temperature' and 'Fluid composition'. Temperature contributes in three distinct fashions during metamorphic transformations:

- It helps in developing straight polygonal grain margins through atomic adjustment within minerals to form low energy recrystallised mosaic texture.
- Coalescence of strain free sub-grains at higher temperature leads to formation of coarse mineral grains with low surface area: volume ratio.
- Changed thermodynamic parameters (viz. entropy and enthalpy) of constituent minerals under elevated temperature lead to destabilization of pre-metamorphic mineral assemblages and ultimate formation of a stable mineral association through mineral reactions. Devolatilization reactions during increased thermal condition ultimately dried up the rock and help in stabilizing less hydrous or anhydrous minerals in place of more hydrous ones.

Pressure also plays some critical role during metamorphism. Pressure of metamorphism can be classified into two categories:

- Lithostatic pressure (P_{litho}) or confining pressure.
- Directive pressure ($P_{directive}$) or deviatoric stress.

Lithostatic pressure(P_{litho}) exerted by rock column during burial is dependent upon density (ρ) of overlying rock and the depth (Z) and is uniform in all directions like hydrostatic pressure under water, and can be expressed as $P_{litho} = \rho. Z. g$ (g = accelaretion due to gravity). Increased lithostatic pressure always leads to decrease in volume of the concerned rock and also helps in stabilization of minerals having more dense atomic structure through polymorphic transformation of mineral of higher molar volume by a denser one. Deviatoric stress, on the other hand, is not uniform in all directions and is invariably associated with tectonic activity during metamorphism. This directive pressure is actually instrumental in modifying the structure of the rock through deformation and development of features like foliation, fold or boudin structure in metamorphic rocks.

Fluids residing in pore spaces between grains sometimes play a pivotal role during metamorphic transformations. Fluid augments inter-crystalline diffusive mass transportation of dissolved ions at elevated P-T condition and promoted metamorphic reactions to occur, as the diffusion rate and consequent reaction rate happen to increase exponentially with increasing T (Reaction Rate = $D.\exp(-E/RT)$ where D is Reaction Constant, E= Activation Energy barrier, R=Gas constant and T= temperature in Kelvin).

Limits of Metamorphism

The initiation of metamorphism begins beyond diagenesis of sedimentary rocks. Diagenesis is the process of incipient transformation of sediments during burial through which they lithified and gradually changed to sedimentary rocks within a maximum temperature range of 200 °C and 30 ° MPa (Mega Pascal) pressure corresponding to ~ 3 Km of crustal depth. Metamorphism, therefore, starts to operate at T > 150° +/−50 °C and P >300MPa.

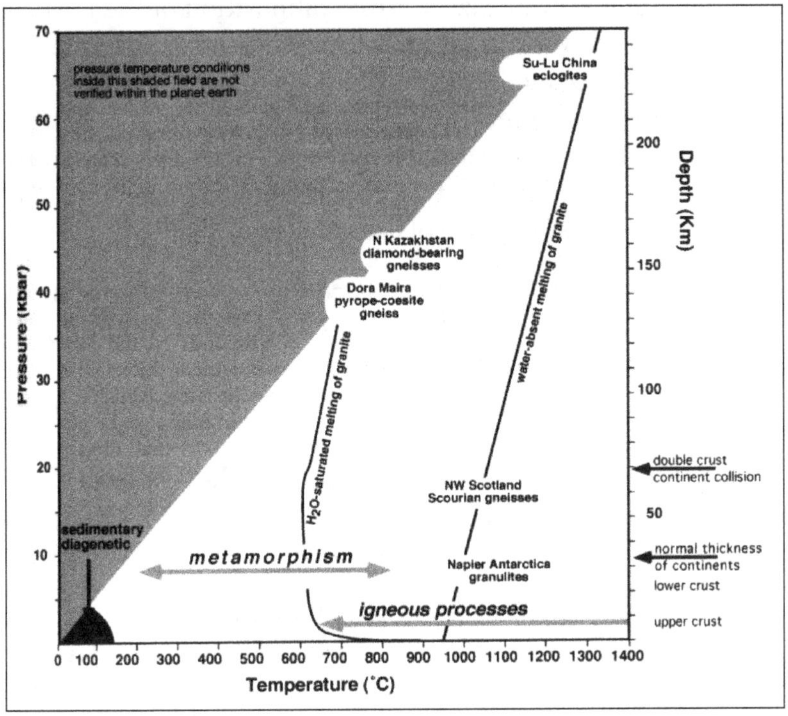

The upper limit of temperature of metamorphism, on the other hand, corresponds to the solidus temperature of common crustal rocks where onset of partial melting takes place at T ~ 850°+/−50 °C. Beyond this the principles of molten rocks of igneous petrology starts to operate. The upper pressure limit of metamorphism is approximately ~60Kbaar which corresponds to ~200 Km depth. Geological processes like ponding of magma at the base or inside the crust, subduction of lithosphere, continent-continent collision, crustal scale tectonics leading to folding and thrusting etc. are responsible for changing the temperature and pressure inside the crust. The time dependent interplay of tectonics and the response of the heat flow mechnisms inside the crustal segment through conduction, advection or radioactive heating, perturbed the steady state geotherm (dT/dZ = 0) to a transient geotherm and causes the metamorphic transformations to occur.

Major Types of Metamorphism

Metamorphic transformations can be broadly classified into two categories on the basis of aerial extent and presence or absence of deformational features:

- Local metamorphism,
- Regional Metamorphism.

In metamorphism of local extent, the metamorphic changes of country rocks are characteristically localized around an igneous pluton that has intruded the rock. The high temperature available during cooling of the igneous intrusive is responsible for elevating the thermal budget of the country rock and promotes metamorphism. The intensity of change in mineralogy, grain size and/or composition of the host rock gradually decreases away from the pluton with diminishing temperature difference and develop a metamorphic aureole of Contact Metamorphism within the country rock around the intrusive body. Since this type of metamorphic transformation is only dependent upon heat transfer due to localized temperature difference and deviatoric stress does not need to play any vital role, the rocks of contact metamorphism show a characteristic haphazardly oriented hornfelsic texture of coarse grains.

Regional metamorphism, on the other hand, is pervasive within large aerial extent and is not confined around any igneous intrusions. Orogenic metamorphism is the most important variety of Regional metamorphism where compressional tectonic activities in destructive plate margins develop highly deformed metamorphic rocks with prominent foliation. In regional metamorphism deviatoric sress uses to play important roles along with elevated temperature and $P_{lithostatic}$ in shaping the metamorphic changes. Foliated metamorphic rocks with schistose or gneissose texture are the products of orogenic metamorphism.

One of the most important emphasis point of metamorphic study is to estimate the T and P of peak metamorphic equilibration and have an idea of the transient geothermal gradient with an attempt to formulate a tectonic model responsible for perturbing the thermal structure of the crust and triggering the metamorphism. Metamorphic Petrologists tried to address this issue of segmenting the P-T range of metamorphism during tectonic disterbances for a long time and gradually evolve the concepts of Metamorphic grade, Metamorphic depth zones and Isograd zonation of metamorphic terranes before finally develop the 'Metamorphic Facies 'and 'Facies series' concepts.

Metamorphic Grade, Metamorphic Zones and Concept of Isograd Mapping

Metamorphism, whether local or regional in extent, use to occur from a very low P-T condition near the earth surface just beyond diagenesis, up to a considerably higher thermal and baric environment where partial melting of common crustal rocks take place leading to the formation of migmatites. One of the main objectives of metamorphic petrology is to ascertain a specific P-T window of equilibration for the metamorphic rocks of interest and its variation in space and time and to build up a comprehensive tectono-thermal evolutionary history for the rock. For this purpose an imprecise general term of 'metamorphic grade' was introduced by Tilley to signify the relative intensity of metamorphism or, more precisely the peak temperature of recrystallisation. In field it's very much convenient to arrange the metamorphic rocks into 'Very-low grade', 'Low grade', 'Medium grade', 'High grade' and 'Very high grade metamorphism' segments according to increasing temperature of equilibration. The boundaries between each segment are characterized by a drastic change in mineral composition or mineral assemblage appropriate for the bulk composition of the rock. During prograde metamorphism a mineral assemblage characteristic of low grade metamorphism always transformed to medium grade or high grade under increasing temperature condition while during retrograde metamorphism under decreasing temperature the high grade metamorphic assemblages tend to convert to medium or low grade one. Grade classification of metamorphism, however, gives only a crude idea of relative temperature range and no precise P-T estimation is possible as the boundary between different grades can shift with variation in bulk chemical composition.

A comparatively better scheme for subdividing the P-T space of interest is through zonal mapping of the metamorphic terrane. Each metamorphic grade can be subdivided into one or more metamorphic zones characterized by first-appearance of certain metamorphic index mineral under the prevailing physical condition of metamorphism, and the zones are named according to the index minerals. For example, in Barrovian metamorphic sequence of pelitic rock under prograde metamorphism,

- Chlorite is the index mineral for 'chlorite-zone' under 'very low grade of metamorphism'.
- Biotite is the index mineral for 'biotite-zone' under 'Low grade of metamorphism'.
- Garnet and staurolite are the index minerals for 'garnet-zone' and 'staurolite zone under' 'medium grade metamorphism'.
- "Kyanite' and 'Sillimanite are the index minerasl for 'Kyanite zone' and 'Sillimanite zone' of 'high grade metamorphism'.

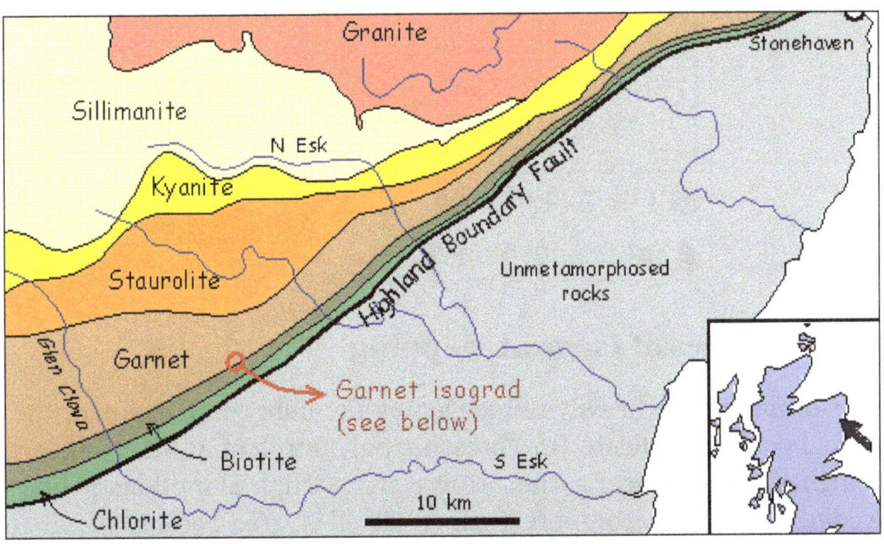

Metamorphic mineral zones in NE Scotland, after Borrow and Tilley.

This zonal sequence suggests that index minerals viz. chlorite-biotite-garnet-staurolitekyanite-sillimanite gradually appear in a barrovian sequence with increasing grade of metamorphism.

The changeover of one metamorphic zone to other with increasing or decreasing P-T condition of metamorphism is always marked by the first appearance or demise of specific index mineral characteristic of that metamorphic zone and this transformation is mainly controlled by some specific metamorphic reaction. The line of first appearance of an index mineral in a metamorphic rock of certain bulk composition is marked as an 'isograd' on a geological map (viz. Bt- in isograd, Gt- in isograd etc.) which is actually the intersection of the reaction line producing the index mineral in P-T-X space with the plane of the map. Geological demarcation of metamorphic zones by joining points of first appearance of certain index mineral typical of rocks having specific bulk composition (isograd mapping) in field thus gives some preliminary idea for intensity of metamorphism. For example, the 'Gt-in' isograd in rocks of normal pelitic bulk usually marks the transition from 'lower metamorphic grade' to medium metamorphic grade and can be express by the reaction:

Chlorite + Biotite = Garnet + Muscovite + Quartz + H_2O

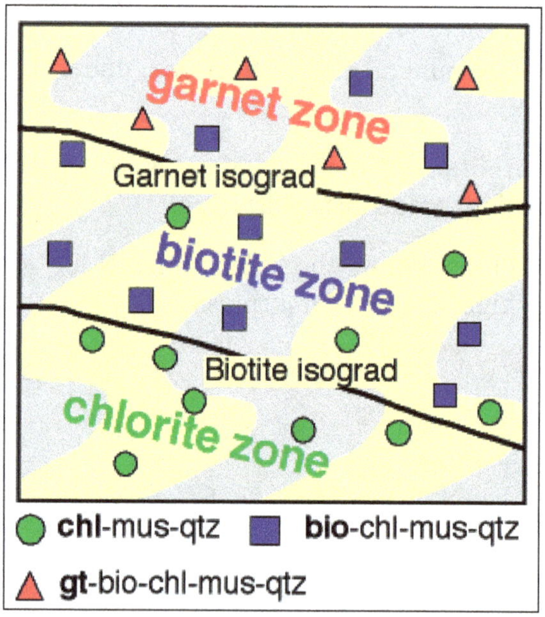

Important Features of Isograd Mapping

- Isigrad mapping is heavily influenced by the bulk composition of the rock. In normal pelite 'andalusite' (low pressure polymorph of Al_2SiO_5) appears after (ie at higher temperature) garnet. However, in 'High aluminous pelitic bulk the 'Andin' isograd is encountered before 'Gt-in' isograd because of enhanced stability of the earlier mineral in Al-rich bulk.

- In some rocks the first appearance of an index mineral is strongly dependent upon the fluid present during metamorphism along with T and P. It's observed that two rocks of identical composition and metamorphosed under same P-T condition often end up with different mineral assemblages depending upon the fluid composition participating metamorphism. In the following reaction in a calcsilicate rock,

  Calcite + Quartz = Wollastonite + CO_2

the first appearence of Wollastonite(Wo- in isograd) at constant T and P is greately suppressed at higher value of XCO_2 that actually enhances the stability of calcite + quartz. So it can be argued that 'isograd 'in field is not actually 'isotherm'.

It's observed that almost all of the index minerals have a wide range of T-P stability and they stably co-exist with other index minerals appearing early or late. This actually proves that 'individual index minerals' are not suitable for precisely constraining the physical condition of metamorphism. Rather it's an assemblage of minerals forming through some reactions are more suitable for understanding the P-T condition of metamorphic transformation.

Origin of Metamorphic Facies Concept

V.M.Goldschmidt's work in Oslo region of southern Norway on hornfelsic rocks of pelitic, calcareous and psammitic composition confirms stabilization of characteristic equilibrium mineral assemblages in different bulk at the inner zone of contact aureoles around granitoid intrusive where T and P of metamorphism are the same for all the rock types. Pentii Eskola while carrying out metamorphic study on identical hornfelsic rocks of Orijarvi region of Finland ascertained the ideas of Goldschmidt that there is a constant relationship between mineralogical composition of a rock and its bulk chemical composition. Eskola also observed stabilization of different mineral assemblages in some of the rocks of Orijarvi region that are identical in chemistry with its Norwegian counterpart studied by Goldschmidt. The 'biotite + muscovite' assemblage of Orijarvi region is characteristically replaced by compositionally equivalent pair of 'k-feldspar + cordierite' in identical pelitic bulk of Oslo region, and Eskola correlates this variation in mineral assemblage due to change in physical condition (T and P) of metamorphism in these two spacially separate regions. This can be explained by the mineral reaction,

$$\underset{\text{(Biotite)}}{2KMg_3AlSi_3O_{10}(OH)_2} + \underset{\text{(Muscovite)}}{6KAl_2AlSi_3O_{10}(OH)_2} + \underset{\text{(Quartz)}}{15\,SiO_2}$$

$$\underset{\text{(Cordierite)}}{3Mg_2Al_4Si_5O_{18}} + \underset{\text{(K-feldspar)}}{8KAlSi_3O_8} + 8H_2O$$

On the basis of this observation Eskola argued that the pelitic rocks of Orijarvi region which are dominated by more hydrous minerals (biotite and muscovite) must have metamorphosed at some lower temperature compared with the pelitic rocks of Oslo region which are dominated by anhydrous minerals like k-feldspar and cordierite.

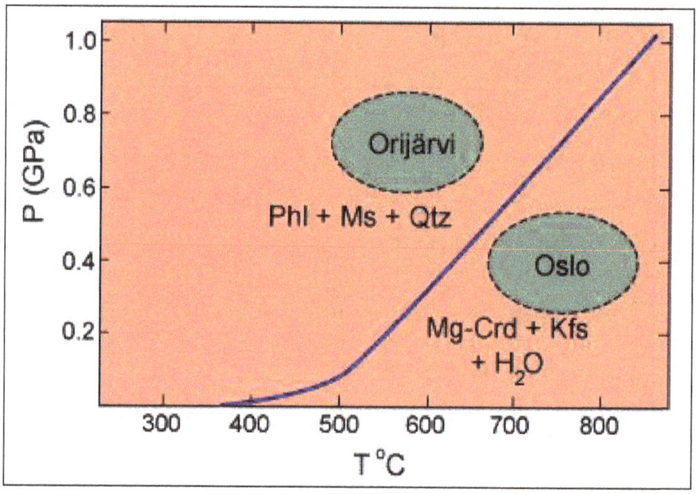

Metamorphic Facies

On the basis of observations stated above and inferences drawn thereafter, the concept of metamorphic facies for qualitatively subdividing the entire P-T range of metamorphism in number of selected domains primarily on the basis of occurrences of stable mineral assemblages developing in rocks of different bulk composition, instead of rely upon appearance or demise of a single index mineral considered in isograd mapping. According to Eskola's facies concept 'any metamorphic rock that has attained chemical equilibrium at a constant P and T condition ,the composition of its mineral assemblages is controlled only by the bulk composition of the rock. With accumulation of precise experimental and thermodynamic data on stability of minerals, an updated metamorphic facies concept was proposed by Fyfe, Turner and Verhoogen which states that "a metamorphic facies is a set of metamorphic mineral assemblages repeatedly associated in space and time so that there is a constant and therefore predictable relationship between mineral composition and bulk chemical composition".

The corollary of metamorphic facies concept is therefore:

- Rocks of identical bulk composition having identical mineral assemblages must have metamorphosed in identical P-T condition of metamorphism and surely belong to same metamorphic facies whatever may be the age and space separation of their occurrences.
- Rocks of a specific bulk chemical composition metamorphosed under different facies condition will definitely produce different mineral assemblages.
- Rocks metamorphosed in same facies condition but showing different mineral assemblages must be of different bulk compositions.

Characteristics of Metamorphic Facies

- Each metamorphic facies is defined by set of mineral assemblages developed in metabasic rocks and are characteristically different from the 'Barrovian Zones' which use index minerals of pelitic bulk.
- The most important feature in 'Metamorphic facies' concept is that it identifies the regularities and consistencies in mineral composition and bulk chemistry of a rock as a function of P and T. The boundaries between different facies coincides with different mineral reactions encountered in metabasic rocks at certain P-T condition. However, a metamorphic facies can only qualitatively constrain a broad P-T range within which a set of metamorphic mineral assemblages are stable and doesn't define the exact P-T condition of metamorphism.
- The facies boundaries are demarcated by certain metamorphic reactions and in majority cases the slope of these reactions are at high angle to the T axis and nearly parallel to P axis. These T-dependent reactions with very high ΔS and low ΔV are mostly devolatilization in nature and hence their positions in P-T space are very much dependent upon composition and quantity of fluid present. Metamorphic facies concept doesn't take the role of fluid composition in fixing the boundary in consideration and hence It is pertinent to say that the boundaries between two consecutive facies are not a fixed line, rather a zone in P-T space.

Classification of Metamorphic Facies

Eskola proposed five metamorphic facies namely Greenschist, Amphibolite Hornfels, Sanidinite and Eclogite and further added three more viz. Epidote-amphibolite, Glaucophane schist and Granulite facies. He also changed the name of Hornfels facies into Pyroxene hornfels facies. Later Coombs added two more sub-greenschist facies viz. Zeolite and Prehnite-Pumpellyite facies and Fyfe proposed Albite-epidote hornfels and Hornblende hornfels facies. The Glaucophane schist facies name later changed to Blueschist facies.

Considering all modifications since inception of facies concept the regional and contact metamorphism can be classified into 11 different facies segments in P-T space. The facies of regional metamorphism includes:

- Zeolite Facies,
- Prehnite-Pumpelleyite Facies,
- Greenschist Facies,
- Amphibolite Facies,
- Granulite Facies,

- Blueschist Facies,
- Eclogite Facies.

The hornfelsic rocks of contact metamorphism can be classified into:

- Albite-epidote hornfelsic Facies,
- Hornblende hornfelsic Facies,
- Pyroxene hornfelsic Facies,
- Sanidinite Facies.

The Epidote-amphibolite facies of Eskola is found to be a transitional sub-facies between Greenschist and Amphibolite and can be encountered only in regional metamorphism along intermediate geothermal gradient and hence not considered as a separate facies.

The following figure shows the relative positioning of metamorphic facies for regional and contact metamorphism in P-T space.

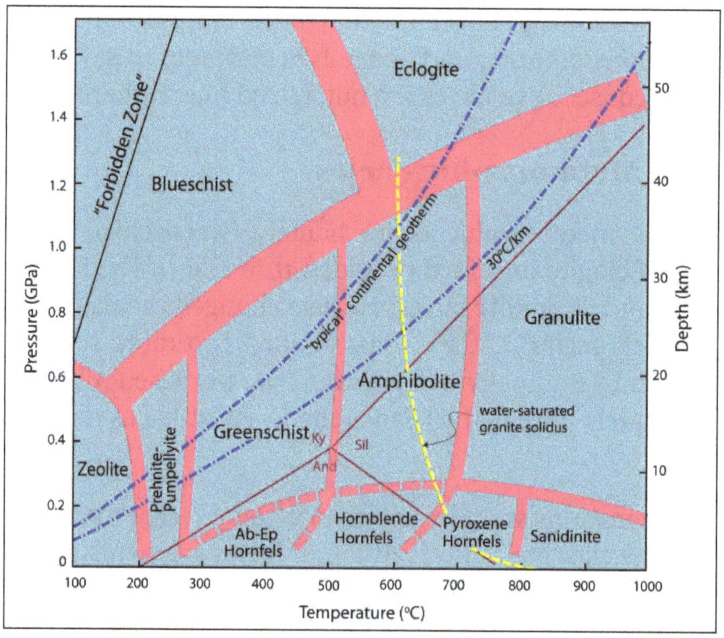

It is evident from the above diagram that a steep geothermal gradient (dT/dP ~ 55 °C/Km) is required for developing hornfelsic rocks of contact metamorphism while regional metamorphism along normal or intermediate geothermal gradient (dT/dP ~ 30 °C/Km) produces rocks of Barrovian sequence ie. from Zeolite - Prehnite-Pumpellyite – Greenschist – Amphibolite to Granulite facies. Regional metamorphism along a shallow geothermal gradient (dT/dP ~ 15 °C/Km), on the other hand, helps in producing rocks of Blueschist and Eclogite facies.

Description of Individual Metamorphic Facies

For the sake of discussion Metamorphic facies has been sub-divide into three categories:

- Metamorphic facies of normal (intermediate) geothermal gradient (dT/dP ~ 25-35 °C/Km).
- Metamorphic facies of shallow geothermal gradient (dT/dP ~ 15-20 °C/Km).
- Metamorphic facies of steep geothermal gradient (dT/dP ~ 50-55 °C/Km).

The first and second subdivisions incorporate the sequence of metamorphic facies developed in Orogenic/Burial regional metamorphism with increasing T or T-P both and can be encountered in rocks of subduction and Continental collisional set up. The subdivision third constitutes the facies developed in contact aureole around igneous intrusions.

Metamorphic Facies of normal (intermediate) geothermal gradient (dT/dP ~ 25-35 °C/Km): The sequence of facies encountered with increasing grade of metamorphism (i.e. with increasing T or T-P both) under intermediate geothermal gradient includes: Zeolite → Prehnite-Pumpellyite → Greenschist → Epidote-amphibolite → Amphibolite → Granulite.

Zeolite Facies

This metamorphic facies is encountered at the lowest P-T condition (~150-300 °C; <3-4 Kb) of regional metamorphism of basic igneous rocks and greywacke and characterized by stabilization of Zeolite group of minerals + quartz + albite +clay minerals. The Zeolites are hydrated Ca-Na- Al silicate and include minerals viz.

Analcime ($NaAlSi_2O_6.H_2O$), Laumontite ($CaAl_2Si_4O_{12}.4H_2O$), Heulandite ($CaAl_2Si_7O_{18}.6H_2O$), Stilbite($CaAl_2Si_7O_{18}.7H_2O$), Wairakite($CaAl_2Si_4O_{12}.2H_2O$), Lawsonite ($CaAl_2Si_2O_7(OH)_2.H_2O$) Zeolite facies is untraceable in pelitic bulk because of lack of Ca in it. The characteristic zeolite facies mineral assemblage develops in meta-basic bulk in presence of little or no deviatoric stress during Burial metamorphism and is free from any foliation. The stability of zeolites are very much dependent on XCO_2, as increasing partial pressure of CO_2 stabilizes Clay mineral + calcite in place of zeolites.

Prehnite-Pumpellyite Facies

This is another sub-greenschist facies assemblage beside zeolite that encountered in meta-basic rock of Burial metamorphism. The facies is characterized by the presence of prehnite + pumpellyite with chlorite, calcite +/- zeolite. Prehnite is a Ca-Al silicate

($Ca_2Al_2Si_3O_{10}(OH)_2$) and pumpeyllite contains some Fe, Mg along with Ca, and Al ($Ca_4(Mg, Fe)Al_5Si_6O_{23}(OH)_3.2H_2O$). Wairakite and laumontite of Zeolite facies breaks down to form prehnite around 3 Kb at 300-400 °C to mark the onset of Prehnite-pumpellyite facies in basic bulk. This facies is also not traceable in pelitic bulk composition.

Greenschist Facies

The transition from Prehnite-pumpellyite facies to Greenschist facies is marked by the breakdown of prehnie +/- zeolite to zoisite.at ~ 400 °C. The typical mineral assemblage of Greenschist facies:,

- In metabasic rock: chlorite + albite + zoisite + quartz + actinolite + epidote +/- calcite +/- sphene +/- ilmenite.
- In metapelites: chlorite + muscovite + biotite +/- albite +/- chloritoid +/- Al_2SiO_5
- In metagreywacke: albite + epidote + muscovite + chlorite + quartz +/- biotite
- In calc-silicate rock: dolomite + quartz + talc +/- tremolite + calcite

The facies name came from the presence of green coloured minerals viz. chlorite, epidote/zoisite and actinolite in profusion in metabasic rocks. The stability of Actinolite is however strongly dependent upon factors like value of PCO_2 and Fe/Mg ratio in the bulk besides P and T. High PCO_2 contributes to the higher stability of chlorite + calcite in place of actinolite ,while a higher Fe/Mg value in bulk prefers early stabilization of actionolite. In low Ca-high Mg metabasic bulk a different amphibole 'Cummingtonite ($Mg_7Si_8 O_{22}(OH)_2$) appears in greenschist facies condition instead of actinolite.

Greenschist-Amphibolite Transition Facies

The characteristic mineral assemblage of greenschist facies in metabasic rock records some remarkable changes in this transition facies before move on to Amphibolite facies condition. The important mineralogical changes encountered in this facies includes:

- In metabasic rock: Plagioclase composition changes from Albite to Oligoclase. In Plagioclase solid-solution series there is a pronounced miscibility gap between albite and oligoclas which is called the 'peristerite gap'. In this transition zone from greenschist to amphibolite facies the plagioclase composition marks a drastic change from An3 to An18 because of this peristerite jump. Amphiboles of greenschist facies become more Al-rich through pergasitic and tscermakitic substitutions.

- Chlorite becomes more Mg-rich and eventually disappears.
- Epidote decreases in abundance and eventually disappears.

The characteristic mineral assemblage : epidote + pargasitic hornblende + oligoclase + /- chlorite:

- In pelitic bulk : Garnet (almandine-pyrope solid-solution) makes first appearance.

  Assemblage : biotite + muscovite + chlorite + garnet + plagioclase +/- chloritoid
- In calc-silicate rock: dolomite + tremolite + quartz +/- talc +/- calcite

Amphibolite Facies

This metamorphic facies marks some characteristic changes in rocks of different bulk composition at temperature ~ 550 °C.

- In metabasic rock: Plagioclase composition becomes more anorthitic.

In greenschist facies at 475 °C plagioclase is approximately An_5 where as in the amphibolites facies around 550 °C plagioclase is ~ An_{45} (andesine). Hornblend is the stable amphibole, Garnet may appear while epidote becomes insignificant. Assemblage: hornblende + andesine +/- garnet +/- epidote.

- In pelitic bulk: Staurolite + Al_2SiO_5 polymorph (kyanite / sillimanite) make their appearance.Chlorite disappears. Assemblage: staurolite + garnet + muscovite + biotite + kyanite/sillimanite + quartz + albite.
- In calc-silicate bulk: dolomite + diopside + quartz +/- forsterite +/- grossular garnet.

Changes in Amphibolite-Granulite Facies Transition

The change over from amphibolite to granulite facies through this transition zone marks some characteristic mineralogical transformation.

(a) In metabasic rock: clinopyroxene stabilizes along with hornblende + plagioclase +/- titanite +/- sphene. No epidote, no chlorite.

(b) In pelitic bulk: staurolite breaks down. Complete disappearance of muscovite via the reaction, Muscovite + quartz +/- albite = k-feldspar + sillimanite + H_2O/ melt. This reaction marks the first appearance of K-feldspar in pelitic bulk and known as 'sillimanite + orthoclase isograd' or 'second sillimanite isograd'.

Granulite Facies

This facies includes the maximum temperature of metamorphic transformation and formation of anhydrous mineral assemblages during orogenic regional metamorphism. The onset of granulite facies metamorphism is marked by dehydration melting reactions and the temperature must have crossed the wet solidus of crustal rocks. The mineral assemblages are:

- In metabasic rock: orthopyroxene + clinopyroxene + plagioclase + /- hornblende +/-garnet +/- biotite +/- quartz +/- rutile +/- ilmenite
- In pelitic bulk :perthitic K-feldspar + sillimanite + garnet + hypersthenes + albite + hercynite + quartz
- In calc-silicate rock : scapolite + wollastonite + grossular garnet + plagioclase + sphene +/- quartz +/- calcite

Metamorphic Facies of Shallow Geothermal Gradient (dT/dP ~ 15-20 °C/Km)

The series of metamorphic facies encountered in this type of orogenic regional metamorphism with increasing T and P are, Zeolite→ prehnite-pumpellyite→ greenschist→ blueschist → eclogite. In some cases the geothermal gradient is so shallow that rocks may directly move to Blueschist facies from Prehnite-pumpellyite without transgressing to greenschist facies. Blueschist and Eclogite are the two metamorphic facies where highpressure-low to moderate temperature mineral assemblages got stabilized in appropriate bulk.

Blueschist Facies

This is the only high-pressure-low temperature metamorphic facies of regional metamorphism. The diagnostic minerals of blueschist facies in metabasic bulk include glaucophane, lawsonite, jadeitic pyroxene, phengitic muscovite, aragonite and pumpellyite. These minerals may occur in combination with minerals such as actinolite, chlorite, albite, epidote, quartz, paragonite and stilpnomelane.

- Prehnite-pumpellyite to Blueschist facies transition assemblage: Glaucophane + lawsonite + albite + chlorite +/- aragonite.
- Greenschist to Blueschist facies transition assemblage: Glaucophane + zoisite/ epidote + albite + actinolite +/ chlorite.

Eclogite Facies

This is the 'high-pressure-moderate to high temperature' facies of regional metamorphism. The eclogite facies is characterized by the garnet (pyrope-grossular rich) + omphactic pyroxene association in metabasic bulk and complete absence of plagioclase of any composition. The other minerals include kyanite, quartz and rutile.

In metasediments: talc + kyanite + phengitic muscovite (whiteschist).

- In high pressure-moderate temperature eclogite facies lawsonite and epidote are stable along with glaucophenic amphibole + garnet. This assemblage marks the transition from Blueschist to Eclogite.
- In high pressure-high temperature eclogite facies omphacite + pyrope rich garnet are the major minerals in metabasic rock.This assemblage is typical of Granulite to Eclogite facies transition.

Metamorphic Facies of Steep Geothermal Gradient (dT/dP ~ 50-55 °C/Km)

The sequence of metamorphic facies developed in steep geothermal gradient includes mineral assemblages of contact metamorphism with hornfelsic texture. The facies sequence from outside towards central part of aureole includes,

Albite-epidote hornfels → hornblende hornfels → pyroxene hornfels → sanidinite facies.

The mineral assemblages of albite-epidote hornfels are same as that of low pressure part of greenschist facies, minerals developed in hornblende hornfels facies are similar to that of amphibolite facies while pyroxene hornfels assemblage matches with that of amphibolite-granulite transition zone.

Sanidinite facies is very rare and can only be observed very close to the contact of a hot basic intrusive (T~ 650-850 °C; P~ very low). In this facies the country rock sometimes show local melting and mark the stabilization of sanidine, the highest temperature polymorph of kfeldspar.

The mineral assemblages of sanidinite facies include:

- In basic rock : plagioclase + hypersthenes + augite + tridymite.
- In pelitic bulk : sanidine + cordierite + spinel + hypersthenes + sillimanite.

In the following figure the relationship between different regional and contact metamorphic facies and grade of metamorphism are shown.

Many of the regional metamorphic complexes of the world show a gradual variation in their peak condition of metamorphism in space and time and indicate progressive metamorphic changes along some definite field P-T curve known as 'Metamorphic Field Gradient (MFG)'.Initially it was believed that all the metamorphic terranes of the world showing progressive metamorphism must have MFG akin to Barrovian Metamorphism. However, on the basis of petrographic studies on the rocks of Japan, opined that the progressively metamorphosed regional metamorphic terranes of the world are actually having diverse mineralogical assemblages depending upon variation in baric condition of metamorphism. The three polymorphs of Al_2SiO_5 (Andalusite,the low pressure polymorph; Sillimanite,the high temperature polymorph and Kyanite,the high pressure polymorph) are very common in middle to higher grade zones of many progreesively metamorphosed terranes and are stable over a broad temperature and pressure range of metamorphism. On the basis of their stable occurrences, Miyashiro suggests that each progressively metamorphosed terranes can develop a characteristic sequence of metamorphic facies assemblages due to variation in P/T ratio along some field P-T curve and he named it as 'Metamorphic Facies Series".

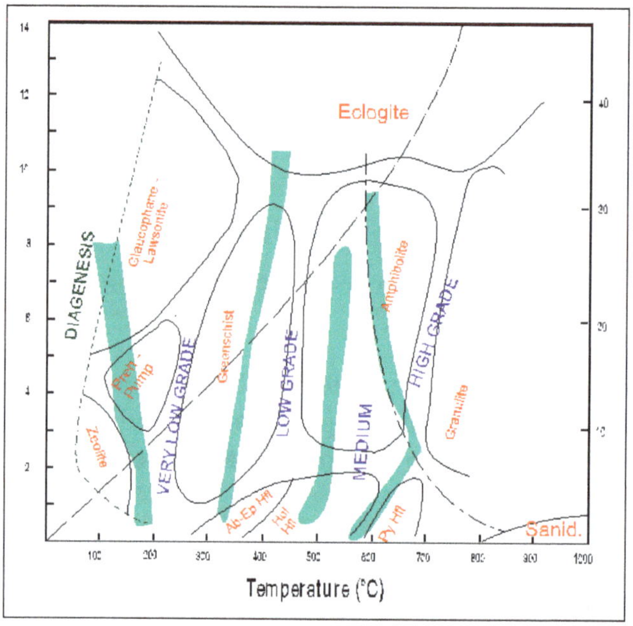

Types of Metamorphic Facies Series

All the regional metamorphic complexes showing progressive metamorphism have been classified into three (3) metamorphic facies series on the basis of positioning of the Metamorphic Field Gradient in P-T space. MFG actually marks the variation of peak P-T condition on the field surface along a line with positive slope which is perpendicular to the thermal peak isotherm or dehydration-reaction isograd.

The three metamorphic facies series classified with reference to the stability field of Al_2SiO_5 polymorphs, glaucophanic amphibole and jadeite + quartz are as follows:

Low P/T Type Facies Series

- In this metamorphic facies series andalusite is stable at lower temperatures and sillimanite appears in higher temperatures.
- Example: Abukuma - Ryoke metamorphic belt, Japan; The Buchan region of Scottish Highlands.

- In these belts ~ 700 °C temperature has been reached at a shallow depth of 8-11 km (~ 3-3.5 Kb pressure).

Medium P/T Type Facies Series

- In this type of metamorphic facies series kyanite is stable at lower temperatures while sillimanite appears in higher temperature domains.
- Example: Barrovian region of Scottish Highlands.
- In this facies series ~800-850 °C of metamorphism reached at a pressure of 7.5-8.5 Kb pressure.

High P/T Type Facies Series

This metamorphic facies series is characterized by the presence of glaucophanic amphibole or glaucophane + jadeite in basic bulk compositions. Glaucophane is a Na-amphibole having Al in octahedral site and stable in high pressure- low to moderate temperature condition while Jadeite is a Na-Al-bearing clinopyroxene having higher pressure stability compared to glaucophane.

This facies series can be divided into two sub-groups depending upon relative occurrence of high pressure minerals:

(i) High pressure transitional type: This type is characterized by presence of glaucophane only and no jadeite is present. The pressure condition is in between kyanite-sillimanite type and jadeite-glaucophane type.

(ii) Jadeite - glaucophane type: This type of facies series occurs at pressures where not only glaucophane is stable but also jadeite + quartz assemblage is stable.

The following figure shows the positioning of different metamorphic facies series.

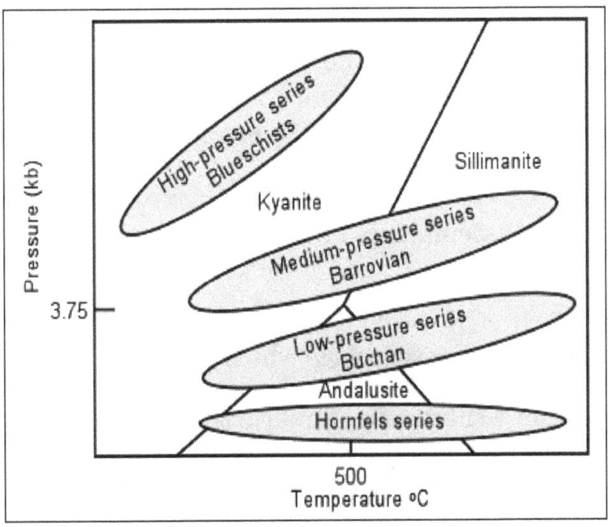

High-temperature Metamorphism

The High-temperature metamorphism is encountered in regional as well as contact metamorphism where the crustal rocks show features of partial melting. The facies of regional metamorphism that includes high-temperature metamorphism are:

- Amphibolite-granulite transition facies,
- Granulite facies,
- Ultra-high temperature segment of granulite facies.

The pelitic and quartzo-feldspathic rocks that undergone dehydration reactions under lower to medium grade of progressive metamorphism up to the P-T conditions of amphibolite facies, first experience dehydration melting reaction after reaching the 'fluidpresent solidus for crustal rocks at the amphibolite-granulite' facies transition zone and this marks the onset of high-grade metamorphism for common crustal rocks. In normal pelitic rock this occurs around 650-675 °C at ~5.5Kb when muscovite breaks down through the reaction:

Muscovite + quartz+ H_2O = Kfeldspar + sillimanite + melt.

This reaction marks the initiation of anatectic migmatitsation at high temperature metamorphic condition.

Granulite facies rocks that form through 'fluid-absent dehydration reactions of biotite and hornblende at still higher temperatures of ~800-875 °C form typical anhydrous minerals viz, orthopyroxene, clinopyroxene, garnet, kfeldspar, sillimanite, kyanite,-cordierite,spinel etc. that constitute rocks of high-temperature metamorphism like charnockite, khondalite, leptynite and mafic granulite.

A sub-group of high-temperature (HT) granulite facies metamorphism is a thermally extreme type crustal condition where peak temperature of metamorphism exceeds 900 °C at lower crustal depth. This is known as 'Ultra-high temperature (UHT) metamorphism. This unusually extreme thermal condition can be recognized either by robust thermometry or by the presence of a diagnostic mineral assemblage in an appropriate bulk composition and oxidation state.

The characteristic mineral assemblages of UHT metamorphism are,

- In pelitic bulk:
    - Sapphirine + quartz assemblage which indicates T >1000° to 1050 °C.
    - Orthopyroxene + sillimanite + quartz assemblage indicates T >900 °C.
    - Osumilite + quartz stability at T> 950 °C at low pressure (6.5-8Kb).

- Spinel(with low Zn content) + quartz assemblage indicates T> 950 °C.

- In calc silicate bulk:
  - Clinopyroxene + scapolite + wollastonite +/- grossular garnet assemblage where scapolite has high 'equivalent An" content.

The facies of high-temperature in contact metamorphism includes sanidinite facies that develops near to the igneous intrusive and often form partially melted rock 'Buchite'.

High-pressure Metamorphism

High-pressure metamorphism is characterized by stabilization of certain minerals through pressure-dependent 'Net-transfer' reactions in appropriate bulk that have low molar volume and high density. Ex: kyanite, glaucophane, jadeite, omphacite and pyrope garnet. The metamorphic facies of regional metamorphism that holds these minerals of highpressure metamorphism includes:

- Blueschist facies,
- Eclogite facies,
- High-pressure segment of Granulite facies,
- Ultra-high pressure segment of Eclogite facies.

Blueschist facies marks the onset of high-pressure - low temperature metamorphism where lawsonite (or pseudomorph after lawsonite) appears along with glaucophane in metabasic rocks. With increasing pressure jadeite appears through the reaction, albite = jadeite + quartz. And jadeite + glaucophane assemblage stabilizes at temperature ~350°-400 °C at a pressure of 2GPa.

Eclogite facies is characteristic of moderate to high temperature- very high pressure metamorphism. At moderate temperature-high pressure eclogite facies condition the characteristic minerals are,

- Glaucophane-epidote association,
- Glaucophane-lawsonite association.

Glaucophane-epidote stabilizes at slightly lower pressure-higher temperature eclogites(~600 °C,2.5GPa) compared to glaucophane-lawsonote (450 °C,3-3.5GPa).

The high pressure-high temperature eclogites (800-900 °C, ~ 2.5 GPa) are characterized by complete absence of plagioclase in Qtz-tholeiite and Ol-tholeiite through the reaction, Plagioclase= garnet + kyanite + quartz.

The other characteristic minerals in this eclogites are omphacite (with high jadeite content) , garnet (rich in pyrope content) and rutile.

The high –pressure granulites are characterized by orthopyroxene-absent assemblage of clinopyroxene + garnet + plagioclase +/- hornblende in mafic granulites and kyanite +/- Mg-chloritoid + talc assemblage in metapelites.

The 'Ultra-high pressure (UHP) metamorphism is a sub-facies of Eclogite facies which is characterized by metamorphism of crustal rocks under pressure conditions that exceed the stability of quartz and is recognized by the presence of coesite or micro-diamond or by equivalent P-T conditions (~600-900 °C and from ~2.6 to >10 GPa) determined by robust thermobarometry.

In the following diagram the high-temperature, ultra-high temperature, high-pressure and ultra-high pressure metamorphic fields are shown.

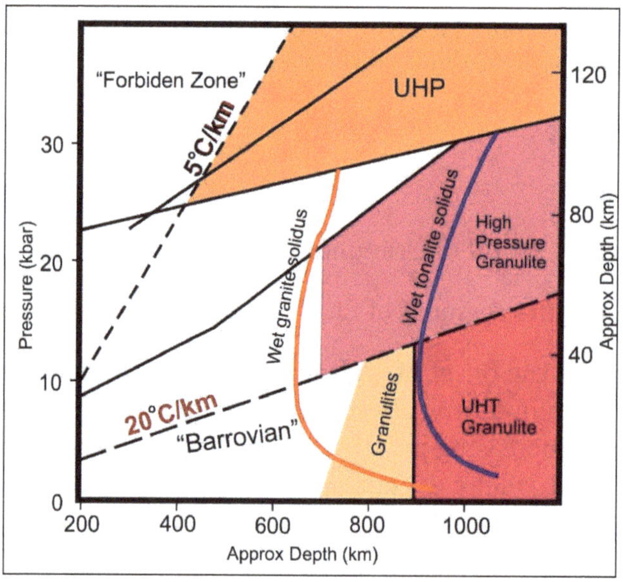

References

- What-is-petrology: worldatlas.com, Retrieved 03, January 2020
- Igneous-petrology: earthsci.org, Retrieved 19, July 2020
- Metamorphic-petrology: earthsci.org, Retrieved 23, April 2020

Chapter 3

Hydrogeology and Engineering Geology

The branch of science that encompasses the scientific study of the distribution, movement and management of groundwater is known as hydrogeology. Engineering geology is an interdisciplinary field that involves the application of geology to engineering particularly geological and geotechnical engineering. This chapter will help the readers in understanding the basic aspects of hydrogeology and engineering geology.

Hydrogeology

It is the comprehensive study of groundwater, its distribution and evolution through time and space, under regional geology. Geohydrology studies water behavior in geological environment, according to hydraulic laws. It includes shaft hydraulic that is drilling to obtain water for different purposes and environment behavior control, according to the kind of drilled material. Both specialties associate regional and local geologic information, superficial and subsoil to contribute optimum groundwater use in favor of sustainable development.

Water and its Importance

Water is the substance that allowed human to establish in specific areas and progress in different tasks. It's the main component of living matter so it constitutes 50 to 90% of the mass of organisms.

The freezing point of water is 0 °C and the boiling point is 100 °C; to atmospheric pressure of 760 mm Hg; and a temperature of 4 °C reaches it maximum density.

Polluted water supply may contribute to the transmission of gastrointestinal diseases, such as cholera, typhoid fever, dysentery and gastroenteritis, as well as viral diseases as infectious hepatitis. Also the lack of water for personal hygiene and environmental sanitation is a contributing factor to the spread of these diseases, so it is of vital importance for all countries to conduct intensive and extensive hydrogeological studies that allow knowing precisely the distribution, quantity and quality of water available in their territorial domains to optimize the sustainable development planning.

Our planet experiences a progressive decrease in the quality and availability of water. A substantial percentage of rural and urban population of the world does not have direct access to clean water. In some regions reserves have been polluted at different intensity levels with toxic chemicals.

Foregoing points out the importance of hydrology, the science that studies water distribution on Earth, its physical and chemical reactions with other substances existing in nature and its relation to life on the planet.

Water perpetual motion between Earth and atmosphere is known as hydrological cycle: water vapor is produced by evaporation at land surface and in water bodies, and transpiration of living beings. This vapor circulates through atmosphere and falls as rain or snow. In condensation and precipitation, rain or snow absorbs from the atmosphere different amounts of carbon dioxide, other gases and sometimes, radioactive substances, as well as small amounts of organic and inorganic material, which is subsequently precipitated along with water.

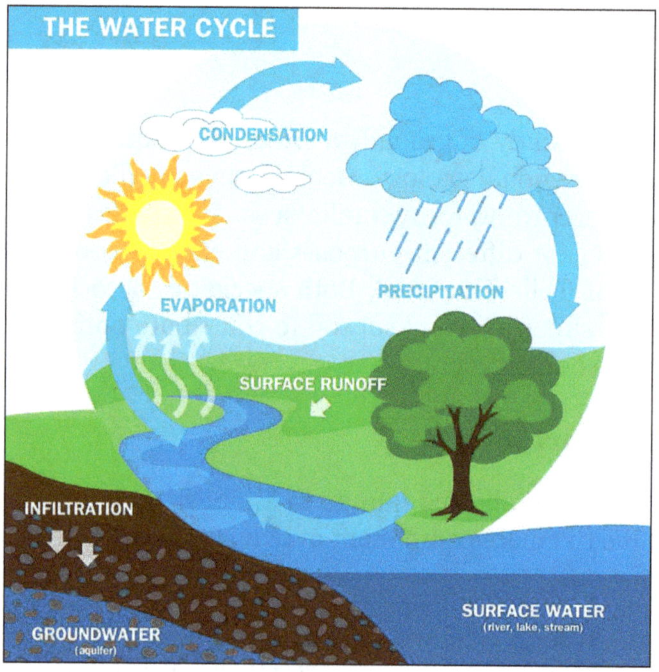

Groundwater

Groundwater is located below the surface, confined to a pressure equal to or greater than atmospheric, saturates the rocky medium through which is moved and stored. Liquid may arise in open spaces of lithological units such as gaps between crystals or grains, failures, joints, lithological contacts and other discontinuities and in rocks cavities caused by cooling, dissolution or weathering.

Movement and groundwater storage depend on the thickness, texture and rocks structure that constitutes the earth's crust, besides the physical ability of water to move through a permeable medium. Also, the imbalance caused by the interaction of different forces, such as gravity and differential pressures, makes the water remains in constant motion, creating deposits which supplies wells and springs, keeping the flow of some streams during periods of drought.

The lithological units that have a greater capacity to store and provide the vital fluid are called aquifers. These may be placed near surface of earth, contained in permeable rocks or confined to deep fractured and/or failed rocks.

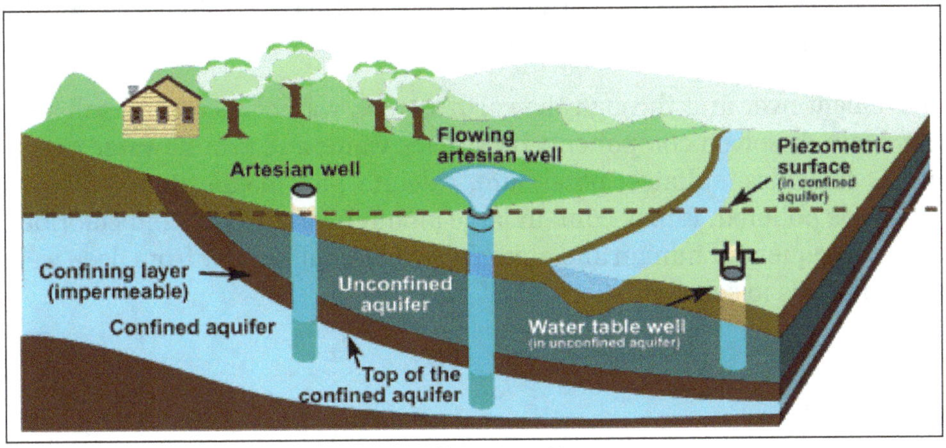

Types of layers aquifers.

Aquifers are supplied or recharged by water that has fallen on the surface of earth as a result of rainfall, hail or snow, and dispersed through soil, sand, gravel and fractured rocks; so water remains stored in one or more geologic formations consisting by permeable rocks which hold back or let liquid pass.

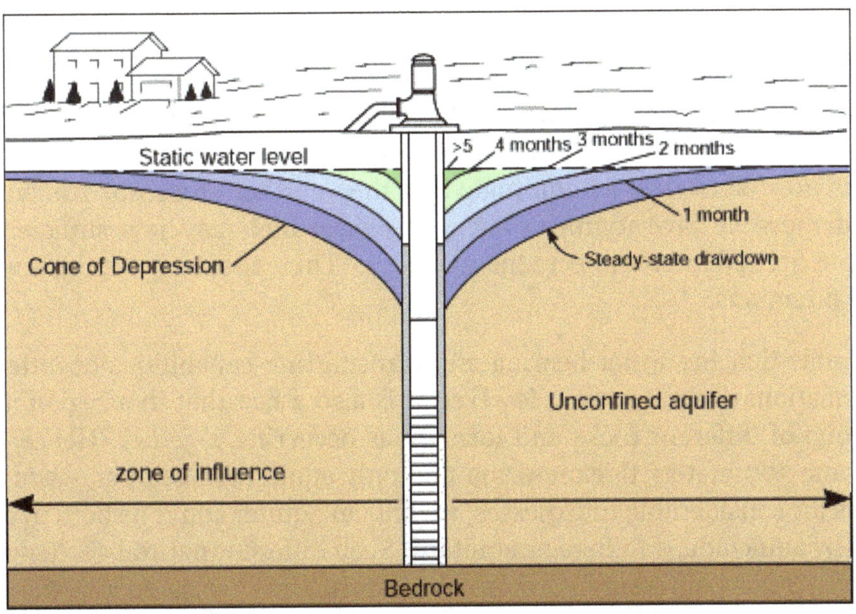

Cone of depression near a well with pump working.

An aquifer of uniform shape and texture and free flowing (either near the surface or in depth) takes the form of an inverted cone, known as the depression cone. This has its apex in the dynamic water level in the well during pumping, and the cone base in the static water level.

Groundwater is found almost everywhere, but can only be extracted in significant amounts when it accumulates abundantly when being trapped in appropriate and specific places of aquifers. It is also present in the upper portion of the soil, where it adheres, by capillary action, to the particles thereof. In this condition, it is called bound water and has different features of free water.

In its movement over and through the earth's crust, water reacts with soil and rocks minerals. Main dissolved components on surface water and groundwater are: sodium carbonates, chlorides, sulfates, potassium, calcium and magnesium. When shallow groundwater is polluted, it may contain large amounts of nitrogen and chlorides compounds as consequent of human and animal waste. Usually water from deep wells contains only minerals in solution.

Aquifer

An aquifer refers to a geological formation, which can store and transmit groundwater in sufficient quantity so that the water can be economically utilized from the aquifer. For this to happen the aquifer material should have reasonably good amount of interconnected pores. None indurated sand and gravels are example of some excellent aquifers. Similarly, aquiclude refers to geological formations like clay, which have good volume of pores, but they are not interconnected. Such formations can only store water but they cannot transmit them for economic yield of water. We also have plutonic rocks like granite, diorite etc. having interlocking granular texture. These rocks do not have much voids/pores, thus negligible interconnection of pores. Such formations are referred to as aquifuge, and they neither store nor transmit water. Besides, all these we have aquitards like clayey silt, which may have sufficient volume of pores, but limited interconnectedness of pores. Thus, they may transmit water only in certain directions.

Now we know that the water bearing and transmitting capabilities of different geological formations are not same. However, it is also a fact that in a region mostly we have a group of different rocks and formations occurring together. Hence, it is rare that for some 300 meters thickness you find only aquifer. In general, when we study formations for considerable thicknesses, we find an aquifer could be both overlain and underlain by aquiclude, aquifuge or aquitard. Such different natural geological setting of an aquifer creates different hydrological regimes. Thus, we categorize the aquifer in to different types.

Types of Aquifer

On the basis of geological settings and distinct hydrological regime, we have mainly four types of aquifer: Unconfined, confined, semi confined and perched aquifer. They are discussed below:

Unconfined Aquifer

An unconfined aquifer is one where a single aquifer body from ground surface is underlain by an impervious layer, which could be either an aquiclude or aquifuge. Thus, here the aquifer can be recharged locally from the rainfall by infiltration and percolation of the rainwater through interconnected pores reaching up to groundwater. The aquifer system is open to surface; air and water are in direct contact through interconnected pores to the water table. A schematic drawing of the aquifer is shown as Figure. It shows a three-layer system, where the top two layers are aquifer underlain by impervious layer. The hydrostatic pressure manifested by water table here coincides with the top of the saturation zone.

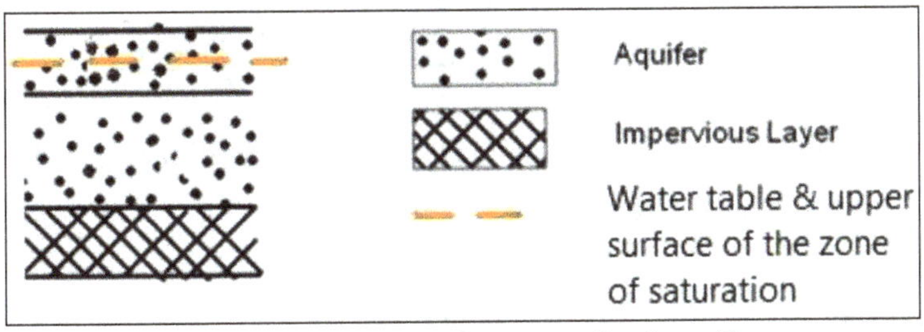

A schematic drawing of an unconfined aquifer.

Confined Aquifer

A confined aquifer is overlain and underlain by impervious layers. This impervious layer could be an aquifuge or aquiclude. As it is overlain by impervious layer, there is no direct local recharge to the aquifer from in situ rainfall. It receives recharge by lateral flow of groundwater from the recharge areas where the aquifer is exposed to surface. The aquifer system is not open to surface locally and the air and water are not in direct contact through interconnected pores to the saturated aquifer locally. This produces a situation that the groundwater pressure is greater than atmospheric pressure. Here the hydrostatic pressure is manifested by an imaginary piezometric surface. The top of the saturation zone coincides with top of the confined aquifer. While the piezometric surface is above the top of the confined aquifer or top of the saturation zone. This also implies that the piezometric surface is also above bottom of the top confining layer.

It is to be noted that on account of overexploitation of groundwater from the confined aquifer, the piezometric surface may declines to the level below bottom of the top confining layer. In such cases, the pressure dynamics of the hydrological system in the confined aquifer shows response similar to that of an unconfined aquifer system.

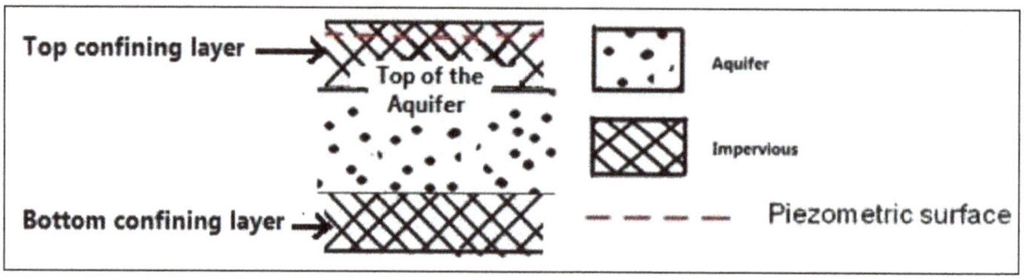

A schematic drawing of a confined aquifer.

Sometimes it is observed that because of topographic setting the piezometric surface is above the ground surface. In such cases, in the wells drilled in confined aquifer water flows naturally out of the wells. Such wells are called as artesian wells.

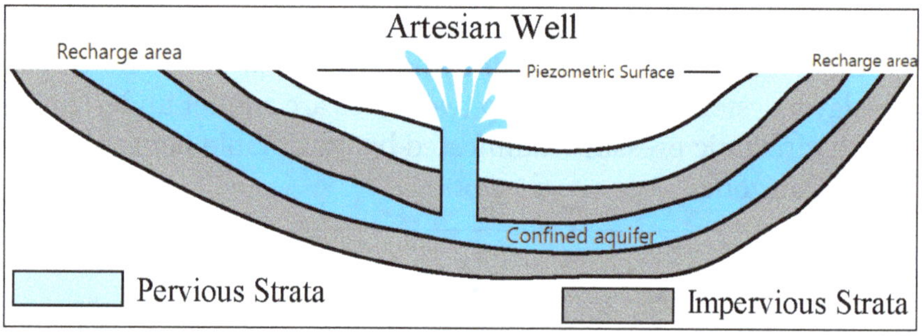

A schematic drawing showing artesian well and recharge area of a confined aquifer.

Semi-confined Aquifer

A semi-confined aquifer is one that is either overlain or underlain by a semi pervious material, which is mostly an aquitard. These aquifers are also referred to as leaky confined aquifers. A schematic drawing of such aquifer is shown as Figure. In an assumed three-layer system, the top layer is semi pervious and the bottom layer is impervious. Such leaky confined aquifers or semi confined aquifers has provision for water movement through the top semi pervious layers.

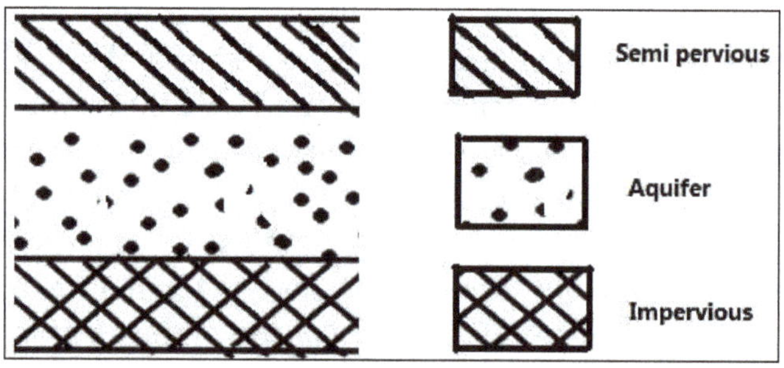

A schematic drawing of a semi confined aquifer.

Perched Aquifer

It often happens that we have a thick unconfined aquifer with some small clay layer at shallow depths. During the process of aquifer saturation through infiltration and percolation by local in-situ recharge, these clay layers create local saturation at shallow depth in the vadose zone much above the water table. Such shallow depth saturation in the vadose zone much above the real saturation zone is often referred to as perched groundwater and the related aquifer is referred as perched aquifer. The perched aquifer along with perched groundwater is shown with a schematic drawing as Figure. It is often seen that shallow tube wells tap the perched aquifers. However, in professional practice it is always advisable to have water wells through the perched aquifer tapping the groundwater below the water table zone.

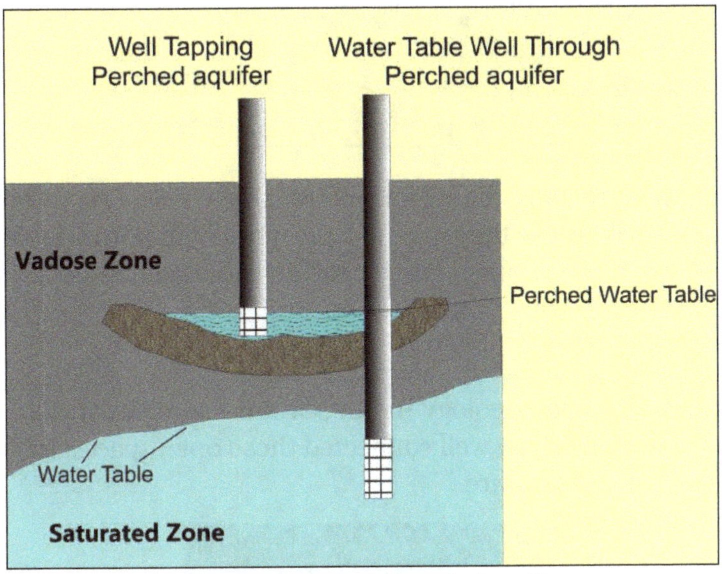

A schematic drawing of perched aquifer.

Properties of Aquifers

Aquifer Materials

Both consolidated and unconsolidated geological materials are important as aquifers. Of the consolidated materials (ie. bedrock), sedimentary rocks are the most important because they tend to have the highest porosities and permeabilities.

Sedimentary rock formations are exposed over approximately 70% of the earth's land surface. These sedimentary formations are typically hundreds to thousands of metres thick, and they are underlain by the igneous and metamorphic rocks that make up the rest of the crust. Although most bedrock aquifers are within sedimentary rock, in some areas igneous or metamorphic rock can be important as aquifers.

Much of the bedrock is also covered with tens to hundreds of metres of unconsolidated sediments (a.k.a. surficial deposits or drift). These include colluvial materials (deposited from mass wasting) alluvial materials (deposited by flowing water), glacial deposits and eolian (wind-blown) deposits. They are "unconsolidated" because they have not been around long enough, and have not been buried deep enough to have become lithified.

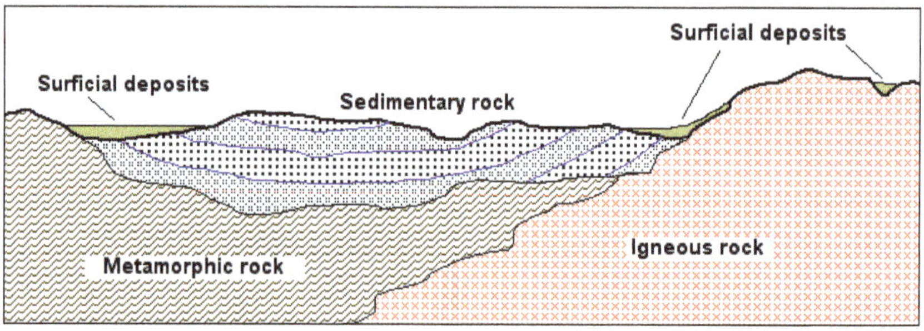

Surficial deposits that are more than a few metres thick can be very important sources of groundwater, partly because they tend to have quite high porosities and permeabilities, and also because they are amenable for the development of wells.

Rock Properties

From the perspective of hydrogeology the important characteristics of rocks are how much open space they have, how well connected those open spaces are, how strong the rocks are, and how soluble they are.

Fractured columnar basalt near Whistler B.C.

Igneous rocks (i.e. rocks formed from the cooling of magma) are comprised of tightly interlocking crystals - primarily crystals of silicate minerals such as quartz, feldspar and amphibole - which tend to be relatively insoluble. In most cases the spaces between crystals are very small. In intrusive igneous rocks these crystals can be quite large (> 1 mm), whereas in volcanic igneous rocks they tend to be much smaller. Some volcanic rocks have vesicular textures caused by the exsolution of gases. Most intrusive igneous rocks (e.g. granitic rocks) are hard and strong, and are less likely to become fractured than other types of rocks. On the other hand, many volcanic rocks are quite well fractured because of their relatively rapid cooling or violent formation.

Sedimentary rocks are formed close to the surface of the earth at relatively low temperatures and pressures. Clastic sedimentary rocks are comprised of weathered and transported fragments of other rocks and minerals. Depending on the degree of sorting and rounding of those fragments, and the extent to which they are cemented together, clastic sedimentary rocks can be quite porous. Some clastic sedimentary rocks are also relatively soft and weak, and are easily susceptible to fracturing. Most sedimentary rocks also have some bedding features that can enhance porosity.

Bedded and cross-bedded Tapeats Fm. Sandstone, Chino Valley, Arizona.

Chemical sedimentary rocks are comprised of minerals made up of material originally derived from other rocks, but transported in solution (such as the Ca^{2+} and HCO_3^- that combine to make calcite). As in igneous rocks, these crystals can be tight and interlocking. On the other hand, some chemical sedimentary rocks, such as limestone and evaporates, can develop significant porosity because of their solubility.

Weathered Quatsino Fm. limestone near to Cowichan Lake, B.C.

Metamorphic rocks form when existing sedimentary or igneous rocks get heated to the extent that the existing minerals start to recrystallize into new minerals. Almost invariably this leads to the development of interlocking textures with low porosities. Most metamorphic rocks are comprised of the relatively insoluble silicate minerals.

Porosity

The empty spaces in between the crystals or fragments that make up a rock represent porosity that can hold water. Porosity is a measure of how much water a body of rock can hold, expressed as a percentage of t rock's volume. The spaces between the grain crystals of a rock are referred to as intergran porosity. Fractures in a rock also represent porosity, and this is known as fracture porosity.

A collection of perfect uniform spheres packed in a rhombic arrangement (as above) will have a porosity of approximately 26%. While sedimentary fragments are never exactly spherical, nor of uniform size, they are also never packed in a perfect arrangement, and porosities of natural sediments can approach or exceed 26%. The porosity of clastic sediments is proportional to sorting and roundness, with the greatest porosities developed in sediments with best sorting (most uniform grain size) and best rounding. Sediments with unrounded and unsorted grains tend to have lower porosities than those that are rounded and sorted, however the more important factor in porosity is the type and extent of matrix and cement.

Clastic sediments with a high proportion of silt and clay-sized matrix (e.g. wacke sandstones) tend t have lower porosities than cleaner sediments (e.g. arenites). Other types of rocks (chemical sediments and igneous and metamorphic rocks), when not fractured, typically have very low porosity.

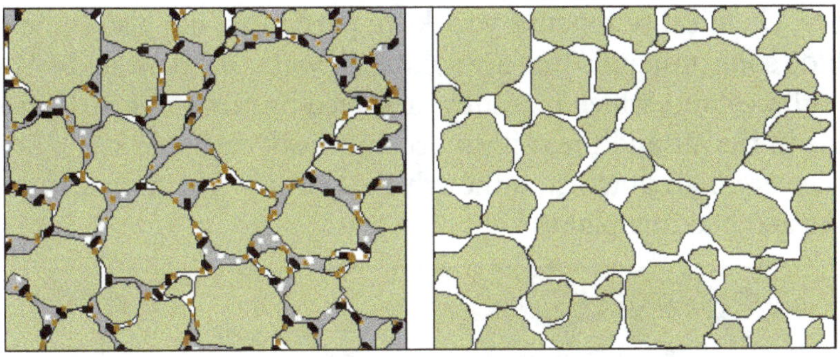

All rocks are susceptible to fracturing, either as a result of tectonic forces, or as part of weathering (e.g. related to pressure release). Strong rocks, like granite or gneiss, may not fracture as easily as weaker rocks, but they can still be sufficiently well broken up to induce significant fracture-related porosity – especially near to surface. Most sedimentary rocks fracture relatively easily. In addition, fracture-like porosity is developed along bedding planes in many cases.

While it is normally possible to estimate or measure the intergranular porosity of a body of rock - using samples of the rock - it can be very difficult to estimate its degree of fracture porosity. Intergranular porosity may be quite consistent over distances of hundreds of metres, but fracture porosity can vary significantly on scales of metres or less. Information on fracture densities and orientations, and on bedding-plane partings, can be derived from outcrops, and then extrapolated to the subsurface, and this type of information can be important to the understanding of groundwater storage and flow potential.

Some of the best aquifers are in rocks that have both intergranular porosity and well-developed fracture porosity.

Rusty coloration indicates water flow along both fractures and coarse-grained layers in Nanaimo Gp. Sandstone.

As noted above, chemical sedimentary rocks in particular, but also some clastic rocks, typically include some minerals that are soluble in water or weak acids. The most obvious example is limestone, which is readily dissolved by rainwater that has reacted with carbon dioxide in the air and the soil (to become weak carbonic acid). Bodies of limestone can have very high porosity, although the porosity is commonly quite localized along fractures and bedding planes.

Dissolution and erosion of calcareous beds in Nanaimo Gp. sandstone, Malaspina Cut.

Specific Yield and Specific Retention

Porosity is a measure of how much water a rock can hold, but it is not necessarily a measure of how much water an aquifer can yield or produce. Firstly, some of the pores may not connect with other pores, or they may be so tenuously connected that water cannot move readily from one to another.

Secondly, while water molecules are very small, and can make their way into even the tiniest spaces in a rock, they are strongly attracted and he to solids by surface tension. The strength of the attraction between the water and the mineral grains is proportional to the distance between the water molecule and the surface. Water molecules within a few microns a surface are very strongly attracted to the surface, while those a few tens or hundreds of microns away are less strongly attracted. If a coarsen grained sediment is allowed to drain (by gravity) much of the water will flow out of it, but some will be retained on the mineral surfaces. In a fine grained sediment, where many of the pores may only be a few microns in diameter, most of the interstitial water will be tightly held to surfaces, and will not flow out.

The proportion of the water in the rock that does drain out readily is known as the specific yield. Fine-grained materials or rock, such as clay or shale, are likely to have specific yields of 5% or less, even where the porosity is in the order of 20 to 30%. Coarse-grained materials or rocks such as coarse sandstone can have specific yields that are closer to their actual porosity – in the range 20 to 35%.

Specific retention is the opposite of specific yield – being the amount of pore water that does not drain readily under gravity. The sum of the specific yield and specific retention is the porosity. A sandstone might have a specific yield of 20% and a specific retention of 6% for a total porosity of 26%. A shale might have a specific yield of 2% and a specific retention of 31%, for a total porosity of 33%.

Hydraulic Conductivity

Hydraulic conductivity also known as permeability is the ability o rock or unconsolidated material to transmit water. Although hydraulic conductivity is partly controlled by intergranular porosity, it is important to remember that very fine grained rocks and materials or rocks with disconnected porosity tend to have low conductivities. Hydraulic conductivity is also controlled by the extent and nature of fracturing and development of bedding-plane partings.

Hydraulic conductivity which is denoted using the symbol K is expressed in velocity units (e.g. cm/s) but it is not a measure of velocity by itself. The hydraulic conductivity can be used to estimate the velocity at which water will flow through rock or sediment, assuming that there is some gravitational force to push the water. Such a force is provided by the hydraulic gradient, or the difference in height of the potentiometric surface (e.g. water table) between two points. The slope of the potentiometric surface is expressed as: dh/dl (change in height over change in length), and flow velocity is estimated using the formula:

$$v = K(dh/dl)$$

In 1856 Henri Darcy estimated hydraulic conductivity by determining the rate of flow of water as a function of the hydraulic gradient, and he carried out a number of measurements using different sediments and varying hydraulic gradients.

The hydraulic conductivity is used to estimate the flow of dilute water a fluid with a viscosity of about 1.1 cp (centipoise) at normal temperatures, and a density very close to 1 g/cm^3. In fact, groundwaters can have quite variable viscosities and densities, due to differing temperatures and dissolved contents. Geologists might also be interested in understanding flow rates of other fluids, such as oil. For this reason, we need to consider the concept of intrinsic permeability - Ki - which is a measure of the properties of the rock or sediment only.

Let us consider the movement of dilute water only, and so we will use the parameter K (and the relationship v = K(dh/dl) to calculate flow rates.

Hydraulic conductivity (or permeability) can be estimated in several ways. For unconsolidated coarse (i.e. sandy) sediments permeability is generally proportional to the grain size.

Permeability can also be estimated using a permeameter in a manner similar to that used by Darcy in 1856. Water, with an applied hydraulic head, is allowed to flow through a sample of rock or sediment, and the rate of flow is measured.

Grain-size data and permeameter readings only provide estimations of the permeability of a sample of the actual material, under conditions that are likely to be very different from those that exist in reality.

Finally, permeability can be measured in the field by observing the rates at which water levels in wells change in response to pump tests (which involve lowering the water level in a well) or slug tests (which involve raising the water level in a well).

Aquifers

As shown on the diagram, geological materials have hydraulic conductivities that can range over many orders of magnitude.

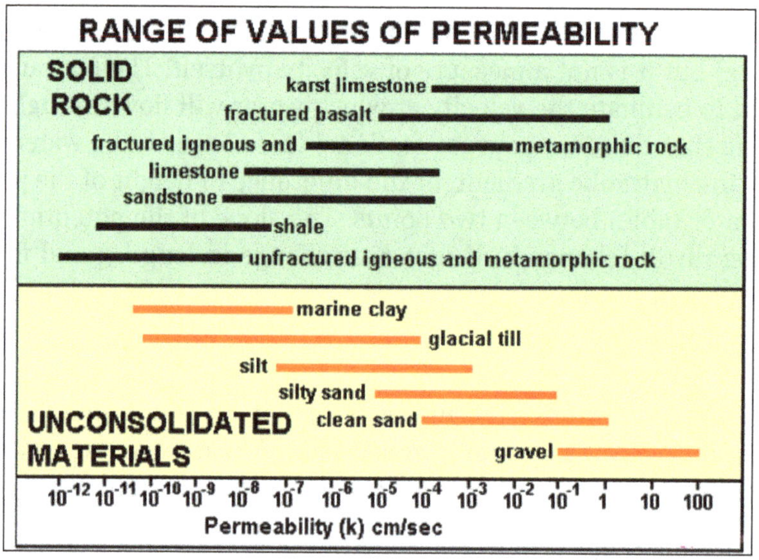

Different rock types can be classified as aquifers or non-aquifers depending primarily on their permeability, but also on the amount of water that is required. An aquifer is any geological unit that can transmit water at a rate that is sufficient to supply a well – although in some cases that may be a very low-flow well. For some purposes, such as for domestic use, the production could be sufficient with a permeability as low as 10^{-5} cm/s, while for others it might have to be much higher. An aquifer may consist of unconsolidated sand or gravel, sandstone, carbonates, basalt or strongly fractured granite or metamorphic rock. The sandstones and mudstones are reasonably good aquifers for domestic wells, but only because they are quite well fractured.

Non-aquifers are generally referred to as: confining layers, aquitards, aquicludes and aquifuges. They might include clay deposits or mudstone, or unfractured igneous or metamorphic rock. Confining layers that are relatively permeable are known as leaky confining layers.

An aquifer in one situation may be considered an aquiclude in another. For example where a siltstone is interbedded with mudrock it is possible that the siltstone could be considered to be an aquifer. The same siltstone interbedded with conglomerate might be thought of as an aquiclude. An aquifuge is supposed to be completely impermeable, but there is no rock that won't let some water through.

- Aquifer: A body of rock that contains sufficient saturated permeable material to yield a significant quantity of water to wells and springs.
- Aquiclude: A body of rock that is incapable of transmitting a significant quantity of water (equivalent to "confining layer").
- Aquitard: Equivalent to acquiclude.
- Aquifuge: (obsolete)A body of rock with no interconnected openings.

An aquifer that is exposed at surface, with no overlying confining layer, is described as an unconfined aquifer. The upper limit of the saturated zone in such a case is known as the water table. The wet but not saturated region above the water table is known as the capillary fringe, while the relatively dry area above that is the vadose zone or unsaturated zone.

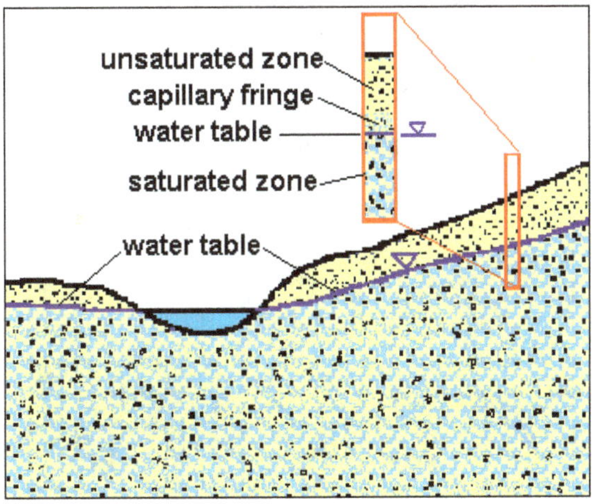

An aquifer that is covered by a confining layer (i.e., a geological unit – either consolidated or unconsolidated - with low permeability) is a confined aquifer. If a well is drilled into a confined aquifer the water in the well may rise above the upper limit of the aquifer. This would be known as an artesian well, and the level to which the water rises defines the potentiometric surface. Under some circumstances the potentiometric surface may ex tend above the ground surface, producing a flowing artesian well.

It is possible that an unconfined aquifer may exist over top of a separate and distinct confined aquifer. In such a case there would be a water table surface and a potentiometric surface in the same area, and they would be different.

If a body of permeable rock has isolated impermeable lenses, such as clay lenses in a sand deposit, some groundwater may accumulate above the clay lenses. The upper limit of such a body of water is known as a perched water table.

The distinction between water table and potentiometric surface is critically important, especially in regional hydrogeological studies where a map of water levels is being prepared in order to understand groundwater flow patterns. For any wells that are being used to measure the water level the hydrogeologist must clearly understand which aquifer is represented by that level, and must be careful not to mix data from different aquifers.

Within an aquifer the water will flow from areas where the elevation of the potentiometric surface is high to areas where the elevation is lower. For example, if the water table is at an elevation of 100 m in well A and 90 m in nearby well B, we can assume that, in general, the groundwater will flow from A towards B. For an unconfined aquifer made up of unconsolidated sediments this principal may apply literally, but there are many situations where the aquifer is either heterogeneous or anisotropic, and the groundwater will be constrained to flow in a specific direction.

Examples of heterogeneous aquifers are shown Figure, and they include aquifers that become thicker or thinner in a certain direction, aquifers that have a change in lithological character, such as a down-dip change grain size or sorting, and aquifers that exhibit distinctive changes in permeability related to sedimentary layering.

Anisotropy within an aquifer can be related to a preferred orientation of grains, as shown in Figure, or to the concentration of faulting, fracturing or bedding planes in a particular orientation.

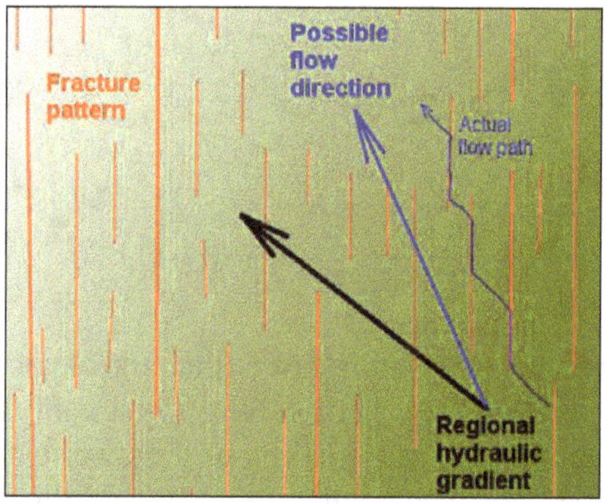

For example, if a unit is strongly and persistently fractured in a north-south direction and the hydraulic gradient is from southeast to northwest (as shown to the left with the heavy black arrow), there could be a tendency for water to be channelled along the fracture planes such that the overall flow direction is towards the north- northwest.

Water Flow within the Unsaturated Zone

Within the unsaturated or vadose zone (above the water table) water does not flow according to Darcy's law. Providing that the material is isotropic and homogeneous, recharging water – such as that from a precipitation event - will flow vertically downward in response to gravity. The rate of flow will be dependant on the hydraulic conductivity of the material, but its rate cannot be estimated using $v = K(dh/dl)$.

Because the distribution of precipitation during a large storm may be variable, and the rates of infiltration through the vadose zone may also be variable, temporary high spots may develop on the water table. This can lead to localized flow systems that are different from the normal flow patterns for that area.

Assessment of Groundwater Quality

The groundwater all over the world has emerged as a potential source for domestic and irrigation purposes. In this case, it becomes important to assess the quality of groundwater before assigning it for domestic or irrigation purposes. We must realize that the natural groundwater without any anthropogenic influences and will also have a baseline concentrating of different dissolved solutes. These solutes are primarily introduced in to the groundwater system by the process of sedimentgroundwater interaction. This natural groundwater quality varies from place to place depending upon nature of soil, nature of subsurface lithology, climatic conditions etc. The dissolved solids in groundwater can be divided in major ions, minor constituents and trace elements. In general, the major ions of groundwater include sodium, potassium, calcium, magnesium cations and bicarbonate, carbonate, chloride and sulphate anions. The minor constituents include small amount of dissolved solids like iron, manganese, fluoride etc. Besides many elements of the periodic table are present in very small concentration in the groundwater and they are referred to as trace elements. Further, at some places we can also find organic compound in groundwater.

Groundwater Quality

When we talk of groundwater chemistry it includes all the dissolved solids in the groundwater system; popularly referred to as the total dissolved solids (TDS), including organic compound and trace elements. While groundwater quality is much more than groundwater chemistry which mainly includes chemical constituents of groundwater. It also includes biotic constituents of groundwater like bacteria, physical parameters like color, turbidity, odor etc. The issue here is which parameters define groundwater quality and what is the standard for these parameters. There is no universal definition. We must understand that water quality is defined in perspective of its desired use. Thus, the water quality desired for drinking purpose will need to be assessed for a set of parameters in comparison to the agreed standard norm. Similarly, the water quality desired for irrigation purpose will need to be assessed for a different set of parameters in comparison to the agreed standard norm for irrigation water. Holistically we can assess the water quality in terms of physical, chemical and bacteriological parameters for different uses.

Groundwater Contamination and Pollution

At this stage, it is desired that we understand the term 'groundwater contamination and pollution'. They are very often used synonymously. Many a times groundwater contamination is referred to geogenic source while pollution is exclusively assigned to anthropogenic source. In literary perspective the word 'contaminate' means making something dirty or impure by adding a substance that may be dangerous, while the word 'pollute' means "adding dirty or harmful substance to land, air, water etc. so that it is no longer pleasant or safe to use". It emerges out from here that loosely these two term terms can be used as synonyms. Since literary meaning of pollution also emphasizes on the fact that during pollution the addition of harmful substance to water, air etc. makes it no longer safe. Therefore, if we want to be more specific we can say that groundwater contamination can be from both natural (geogenic) or human induced (anthropogenic) sources. While groundwater pollution is a stage when groundwater contamination has crossed the safe limits of the desired use. Since historically most of the groundwater pollution was attributed to human activity, in due course of time the use of groundwater pollution was generally restricted to human induced activity leading to occurrence of certain parameters beyond safe limits of the desired use. We must also be very clear here that a groundwater, which is polluted for drinking purposes with respect to certain parameters, might be only contaminated for irrigation uses. Technically the sediment groundwater interaction leading to dissolution of solids and its addition to groundwater below desirable limits for drinking purposes is also an act of groundwater contamination.

Measurement and Reporting of Water Quality Parameters

The water quality parameters are estimated following the standard protocols. The analysis is finally reported by laboratory in unit as given below.

Organoleptic and Physical Parameters

- Colour: Reported in Hazen Unit. Hazen unit basically quantifies colour to a numeric value, which is then used for comparison.
- Odour: It is sensed physically and should be agreeable to a normal human being.
- Taste: Again, it is sensed physically and should be agreeable to a normal human being.
- Turbidity: Reported in Nephelometric Turbidity Unit (NTU). NTU is basically quantification of scattered light at 90 degree from the incident light beam.
- pH: It is measure of negative logarithm to base 10 of the hydrogen ion concentration in moles per liter. It is used for expressing acidity or alkalinity of the solution.

Chemical Parameters

The chemical parameters are mostly reported from a laboratory in term of mg/liter. For pesticides and some trace elements, it can also be represented as micro gram/liter. The interrelationship amongst different units of measurements is:

Milligram per liter (mg/l) ≃ parts per million (ppm).

Mill equivalent per liter (me/l) = mg/l ÷ equivalent weight.

Where Equivalent weight = Formulae weight ÷ charge.

Radioactive Substances

The parameter used for quantifying the concentration of radioactive substances in water is Becquerel/liter (Bq/l). Becquerel quantified radioactivity of a substance.

Bacteriological Parameters

The bacteriological parameters with regards to BIS drinking water standards specifies that the organism shall not be detectable in any 100 ml sample.

Quality Criteria for Irrigation Water

Plants are sensitive to the water quality used for irrigation; the crop yield depends on the type of water used for irrigation purpose. The main parameters identified for assessing water quality for irrigation purposes are: Salinity; Sodium Adsorption Ratio (SAR) and specific ion toxicity (Sodium, chloride, Boron).

Salinity

Very high salinity in irrigation water adversely affects plant growth. The salinity of water is measured by total dissolved solids, generally expressed as mg/l. The total dissolved solids in water control the electrical conductivity of water. Now a days hand held devices are available for estimation of the electrical conductivity of the water. Todd (1980) elucidates and approximates Logan (1961) and Richard's (1954) relationship between TDS and electrical conductivity. It mentions that for water having electrical conductivity in the range of 100 to 5000 micro Siemens/cm the relationship is:

1 mg/l = 1.56 micro Siemens/cm

Since measurement of electrical conductivity is quite easy, most of the time the water quality consideration with respect to salinity is discussed in terms of electrical conductivity. Different agencies and workers have discussed in detail about tolerance of crops vis-à-vis salinity of irrigation water. However, the most holistic consideration with regards to suitability of irrigation water can be taken when salinity is examined vis-à-vis Sodium Adsorption Ratio (SAR). This can be easily done using C-S diagram. The detailed discussion on Sodium Adsorption Ratio and C-S diagram follows further in text.

Sodium Adsorption Ratio (SAR)

The sodium content relative to calcium and magnesium content of the irrigation water controls the rate of infiltration of water to the root zone of crops. It has also been elucidated further by Todd that the soil permeability gets reduced by reaction of sodium with the soil.

$$SAR = \frac{Na}{\sqrt{Ca + Mg} / 2}$$

It is to note here that all the cationic concentrations are in meq/liter.

Specific Ion Toxicity

In this category, we have toxicity primarily by chloride, sodium and boron ions. What happens that circumstantially if the concentration of these ions crosses the critical limit of the crop, the yield reduces drastically? The water needs to be assessed vis –a – vis tolerance of specific crops with regards to concentration of these ions in irrigation water.

Graphical Representation of Groundwater Chemical Quality Data

The chemical quality of groundwater is appreciated better by graphical representation. In the very beginning, we must understand that mostly in graphical representation the major ion concentration (that is Na, K, Mg, Ca, CO_3, HCO_3, SO_4, Cl) is used. Further, while plotting these concentrations are mostly in milli equivalents per liter (meq/liter). The graphical representation of groundwater quality data started by using bar graphs, circular diagram etc. However, limitation of these plotting were that they could not represent substantial number of analysis for one synoptic view. Piper improved upon Hill to produce trilinear diagram for representation of major ion chemistry of groundwater. Besides Richards proposed a diagram for classification of irrigation water. The diagram has Electrical conductivity on x-axis and SAR on y-axis and thus it is also referred to as C-S diagram.

- Trilinear Diagram (Hill-Piper plot): This is one of the most useful diagrams for graphical representation of groundwater quality data. It has unique advantage that large number of data can be plotted on a single diagram. It also helps in ascertaining hydrochemical facies of groundwater. The hydrochemical facies refers to a cation and anion concentration within a distinct zone. Prior to plotting a data, the concentration of major ions is converted in terms of meq/l. Then the cations are tabulated as percentage of total cations and similarly the anions are tabulated as percentage of total anions. The percentage value of each cation is plotted on the cation triangle on the left and percentage value of each anion is plotted on the anion triangle on the right. The data from cation and anion triangles are projected to the central diamond shaped area as shown by dotted lines. In this way, huge number of data can be plotted and the hydrochemical facies variation can be identified.

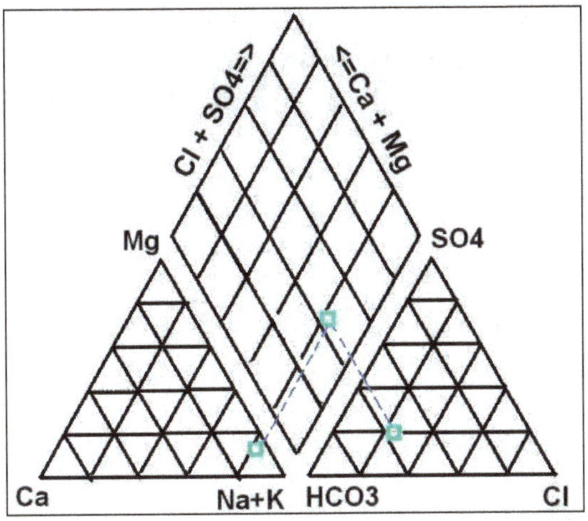

Trilinear diagram for representation of major ion chemistry of groundwater.

- Durov's Diagram: The diagram proposed by Durov has again two cation and anion triangles, and in place of central diamond, area there is a square. Like in a Trilinear plot here again major cations and anions are converted in terms of meq/l and then percentage of each action with respect to total cation and percentage of each anion with respect to total anion is tabulated. These percentage values of the cations and anions are then plotted in their respective triangles. Further, these plots can be projected to central square area as shown by blue line in Figure. In extended Durov's diagram the square can be extrapolated blow and to its right for plotting two more additional parameters.

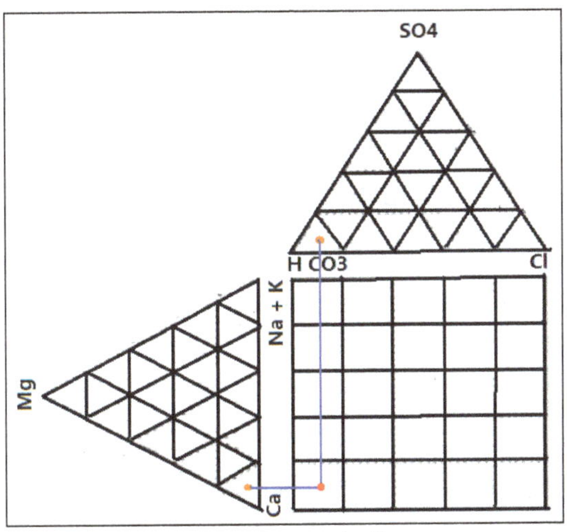

A prototype of Durov's diagram.

- C-S Diagram: The C-S diagram is used for classification of irrigation water. Here on logarithmic x-axis, we plot electrical conductivity in terms of micro Siemens/cm and on arithmetic y-axis, we plot SAR. The x-axis helps in assessing salinity hazard and the y-axis helps in assessing sodium hazard. The whole plot area is divided in to low, medium, high and very high category with respect to sodium and salinity hazard. The location of the sample on plot helps in assessing suitability of water for irrigation purpose. In general, as we move towards higher values of sodium or salinity hazard, the unsuitability of irrigation water increases.

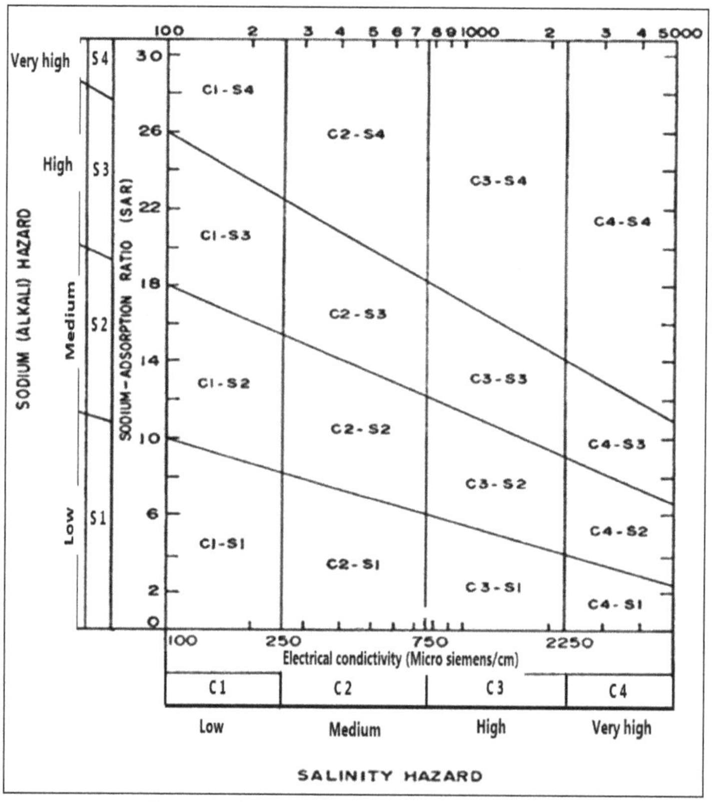

C-S diagram for classification of irrigation water.

Quality Standards for Potable Water

The primary aim of the quality standards for potable drinking water is the protection of public health. The quality of water defined by potable drinking water standards is such that it is suitable for human consumption of health-related chemicals and microbiological constituents, with suitable aesthetic and organoleptic characteristics, and usable for all normal domestic purposes, including personal hygiene.

The WHODWQG is able to provide for the establishment of national standard of drinking water. It describes how guideline values for drinking-water contaminants are to be used, defines the criteria used to select the various chemical, physical, microbiological, and radiological contaminants included in the report, describes the approaches used in deriving guideline values, and presents brief summary statements either supporting the guideline values recommended or explaining why no health-based guideline value is required at the present time. The process for the development of water quality standard is given by showing the protocol of WHO drinking water quality guidelines. Since the regulatory standard should be applicable and feasible to enforce it in each country and area where its natural, geological, cultural and economical situations are different, it is necessary to assess and evaluate the regulatory standard periodically.

Development of National Standards using WHODWQG

The WHODWQG include many drinking contaminants, but also it is unlikely that all of these chemical contaminants will occur in all water supplies or even in all countries. Care should be taken in selecting substances and other characteristics for establishing national drinking water standards. However, a number of factors should be considered, including the types of human activities taking place in the region. For example, if a particular pesticide is not used in the region, it is unlikely to occur in the drinking water. In other cases, such as the disinfection by-products, it may not be necessary to set standards for all substances for which guideline values have been proposed. If chlorination is practiced, the trihalomethanes, of which chloroform is the major component, are likely to be the main disinfection by-products, together with the chlorinated acetic acids in some instances. In many cases, control of trihalomethanes, halo-acetic acids and other DBPs will also provide an adequate measure of control over other chlorination by-products.

In developing national standards, care should also be taken to ensure that scarce resources are not unnecessary diverted to the establishment of standards and monitoring of substances of relatively minor importance. Generally, microbiological indices, inorganic and organic substances, and aesthetic items, are used in national drinking water quality standards.

The potential consequences of microbial contamination are such that its control must always be of paramount importance and must not be compromised. The provision of a safe supply of drinking water depends upon the use of either a protected high quality groundwater or a properly selected and operated series of treatments capable of reducing pathogens to negligible levels, not hazardous to health. Microbial water quality may change rapidly and over a wide range. Short-term peaks in pathogen occurrence may considerably increase disease risks and may also trigger outbreaks of waterborne disease that may affect large numbers of persons. For these reasons reliance cannot be placed on water quality measurements, even when frequent determinations of the safety of drinking water are made. This approach cannot guarantee timely indication of water quality deterioration.

The health risks due to toxic chemicals in drinking water differ form those caused by microbiological contaminants. The fact that chemical contaminants are not normally associated with acute effects places them in a lower priority category than microbiological contaminants, the effects of which are usually acute and widespread.

Therefore, it can be argued that chemical standards for drinking water are of secondary consideration in a supply subject to severe bacterial contamination.

The problems associated with chemical substances in drinking water arise primarily from their ability to cause adverse health effects after prolonged periods of exposure; of particular concern are contaminants that have cumulative toxic properties, such as heavy metals, and substances with carcinogenicity.

The presence of nitrate and nitrite in water, causing both acute and chronic effects, i.e. methaemoglobinaemia (known as blue baby syndrome) and liver ailments, may result from the excessive application of fertilizers or from leaching of wastewater or other organic wastes into surface water and groundwater.

Exposure to a high level of naturally occurring fluoride can lead to mottling of teeth and (in severe cases) skeletal fluorosis and crippling. Similarly, arsenic may occur naturally, and long-term exposure via drinking water may result in various risks to health. In areas with aggressive or acidic waters, the use of lead pipes and fittings or solder can result in elevated lead levels in drinking-water, which may affect the mental development of children.

It should be noted that the use of chemical disinfectants in water treatment usually results in the formation of chemical by-products, some of which are potentially hazardous. However, the risks for health from these by-products are extremely small in comparison with the risks associated with inadequate disinfection, and it is important that disinfection should not be compromised in attempting to control such by-products.

There are few chemical substances in water that can lead to acute health problems except through massive accidental contamination to a water source. Moreover, experience shows that, in such incidents, the water usually becomes undrinkable owing to unacceptable taste, odor, and appearance.

In assessing the quality of drinking water, consumer rely principally upon their senses. Microbial, chemical, and physical water constituents may affect the appearance, odor, or taste of the water, and the consumer will evaluate the quality and acceptability of the water on the basis of these criteria. Water that is highly turbid, highly colored, or that has a detectable level of undesirable taste or odor may be regarded by consumers as unsafe and may be rejected for drinking purposes. In extreme cases, consumers may avoid aesthetically unacceptable but otherwise safe supplies in favor of more pleasant but potentially unsafe sources of drinking water. It is, therefore, vital to maintain a quality of water that is acceptable to the consumer, and the absence of any adverse sensible effects does not guarantee the safety of the water.

Taste and odor problems in drinking water supplies are often the largest single cause of consumer complaints. Changes from the normal taste of a public water supply may signal changes in the quality of the raw water source or deficiencies in the treatment process. It is therefore wise to be aware of consumer perceptions and to take into account both health-related guidelines and aesthetic criteria when assessing drinking-water supplies.

Engineering Geology

Engineering Geology is the application of geology to engineering studies to ensure that the geological factors related to the location, design, construction, operation and maintenance of engineering works are recognized and taken into account.

Engineering Geology provide geological and geotechnical recommendations, analysis and design related to human development and different types of structures. The engineering geologist's realm is essentially about earth-structure interactions or investigating how earth or earth processes impact human-made structures and human activities.

Geological engineering studies can be performed during the planning phases, environmental impact analysis, civil or structural engineering design, value engineering and construction phases of public and private works projects, and post – construction and forensic phases of projects. Geological hazard assessments, geotechnical, material properties, stability of landslides and slopes, erosion, flooding, dewatering, and seismic investigations, etc.

Geological engineering studies are conducted by a geologist or engineering geologist who is educated, trained and has experience in recognizing and interpreting natural processes ; Understanding how these processes affect human – made structures (and vice versa) and knowledge of ways to mitigate hazards caused by adverse natural or human – made conditions. The engineering geologist's main objective is to protect life and property from damage caused by different geological conditions.

The practice of engineering geology is also very closely linked to the practice of geological engineering and geotechnical engineering. If there is a difference in the content of the disciplines, it is mainly the training or experience of the practitioner.

One of the most important roles as an engineering geologist is the study of landforms and earth processes to identify potential geological and associated human-made hazards that may have a significant impact on civil structures and human development. The background in geology provides the engineering geologist with an understanding of how the earth works, which is crucial to mitigating the risks associated with the environment. Many engineering geologists have also graduated with specialized training in soil mechanics, rock mechanics, geotechnics, drainage, hydrology and civil engineering. Such two elements of engineering geologist training provide them with a specific ability to understand and minimize hazards associated with earth-structure interactions.

What is the Importance of Engineering Geology?

The construction of large civil engineering projects requires knowledge of the geology of the area concerned. The geology of an area dictates the location and nature of each of the following structures: Dams, Building foundations, roads and railways. Describe the causes of failure of the slope and possible preventive measures. Discuss a geologist's role in a large civil engineering project's feasibility study and site selection stages.

Engineering Geology helps to ensure a stable and cost-effective model for construction projects. Gathering geological information for a project site is important in the planning, design, and construction phase of an engineering project. Carrying out a detailed geological survey of the area before starting the project would reduce the overall cost of the project. Common fundamental problems in reservoirs, bridges and other buildings are usually directly related to the geology of the region in which they were constructed.

Some civil engineering works require some digging of soils and rocks, and they include the charging of the Earth by building on it. In some cases, excavated rocks may be used as building material, and in others, rocks may form a major part of the finished product, such as a highway or a site f or a dam. The feasibility, planning and design, construction and costing of the project and the safety of the project that depend critically on the geological conditions under which the construction will take place. This is particularly the case in the expanded' greenfield' sites, where the area affected by the project stretches for kilometers over relatively undeveloped land. Sources include the design of the Channel Tunnel and the building of motorways. In the section of the M9 motorway connecting Edinburgh and Stirling, which crosses abandoned oil shale sites, the realignment of the route, on the advice of government geologists, has led to significant savings. For small ventures or those requiring the reconstruction of a limited site, the demands on the geological expertise of the contractor or the need for geological advice will be less, but will never be negligible. In such situations, the site inspection by boring and analyzing samples may be an adequate preliminary to the building.

What Type of Work do Geological Engineers do?

Many of these specialists consult for engineering or environmental firms. Many are employed by departments of the highways, environmental agencies, forest services, and hydro operations.

Construction industries depend on geological engineers to ensure the stability of rock and soil foundations for tunnels, bridges, and highrises. Foundations must withstand earthquakes, landslides, and all other terrestrial phenomena, including permafrost, swamps, and bogs.

Geological engineers are finding better ways for landfill construction and management. They find safer ways to dispose and manage sewage from toxic chemicals and garbage. They plan and design tunnels for excavations.

Groundwater is another specialty of geological engineering. Industries and farms need reliable sources of water, requiring dams or drilling wells at times. These engineers regulate the supply of water to hydroelectric dams; they design dikes and work to prevent shoreline erosion.

Geophysical Exploration Methods

The geophysical investigations are based on the measurements of different physical properties such as density, gravity, seismicity, resistivity, magnetism, radioactivity etc. using specially designed instruments on the ground points as well as in bore holes to sequester information regarding of subsurface. Geophysical methods are rapid, economical and reliable but need calibration from known data or may need borehole data for verification. In fact an appropriate combination of direct and indirect methods not only optimizes the expenditure but also help in getting good results. Apart from geological or engineering geological surveys geophysical explorations are also used for detecting subsurface leakage in pipe lines, sever lines, buried old pipe lines, ammunition storage, abandoned shafts, small mines, and treasures etc. as non geological uses. For some mega civil engineering projects such as site selection of dams, reservoirs, tunnels, bridge foundations etc., geophysical explorations are invariably used to get clear picture of subsurface conditions in following terms:

- Thickness of overburden: Presence of soil or weathered rocks, either to be removed for reaching the sound foundation rocks or to be located as thick loose rock debris to be used as construction material.
- Depth of bed rock: Depth and three dimensional extent i.e. basement relief of sound rocks suitable for foundation.
- Presence of weak zones: Detection of cavities, cavernous rocks, shear and fault zones which may not only act as weak zones but may also act as zones of water leakage as well as major rock structures.
- Groundwater potential: Finding out regions with high groundwater potential and its favourability or non favourability for specific project.
- Rock mass exploration: To understand weathering extent, type, thickness, attitude and extension of different rocks in subsurface.

Gravity Method

This method uses the earth's natural force of gravitation which pulls down different objects towards it at the rate of 980cm/sec^2 or 980 gal (standard value). One gal is an acceleration of one centimetre per second per second. This value varies from equator (978.049 gal) towards poles (983.221 gal) mainly due to variation in the radius of the earth. The theory behind this method is that for every place of the earth there is a theoretical value of 'g' called as 'gT' which is related to the latitude of that place by a formula:

$$g_T = g\,[1 + 0.0052884 \sin^2 \phi - 0.0000059 \sin^2 2\phi]$$

where, 'g' is the standard value of acceleration due to gravity (978.049 gal at sea level) and g_T is the theoretical value of 'g' at a place having latitude ϕ. But if we compare gravity measurements of two places at same latitude we may get different results, primarily due to difference in mean sea levels, the rock mass present between sea level and station level and effect of a topographic high or low near the 'g' measuring station. To get exact results following corrections are to be made:

1. Free air correction: As most of the station are above sea level and 'g' values decreases with increasing altitude. A correction is to be added as +30.86 milligal per 100m increase.

2. Bouguer reduction: The rocks between sea level and station will add to the gravity values depending upon density of rocks and altitude. Hence in this correction some value is to be detected. Free air correction and Bouguer reduction can be worked out for 100m elevation by formula:

$$g_0" = g + H(0.0003086 - 0.0000421\rho) \text{ gals}$$

where "g_0", is gravity corrected for free air and Bouguer corrections, H is the height from the mean sea level and ρ, density of rocks. Terrain correction: The presence of considerable relief near the station has to be taken into account by compartmentalizing the region around the station in radial manner and by taking average height of the compartment as shown in the figure and then calculating its effect as:

$$g_t = k.\rho.\,\phi\left(r_1 + \sqrt{r2^2 + h^2} - \sqrt{r1^2 - h^2} - r^2\right)$$

where g_t gravity after terrain correction, k is gravitational constant, ρ is density of rocks, h, average elevation in centimetres of the compartment above or below the instrument, ϕ angle between radius of inner bounding circle, r_1 and r_2, is of outrebounding circle. Standard charts and tables are available for these corrections.

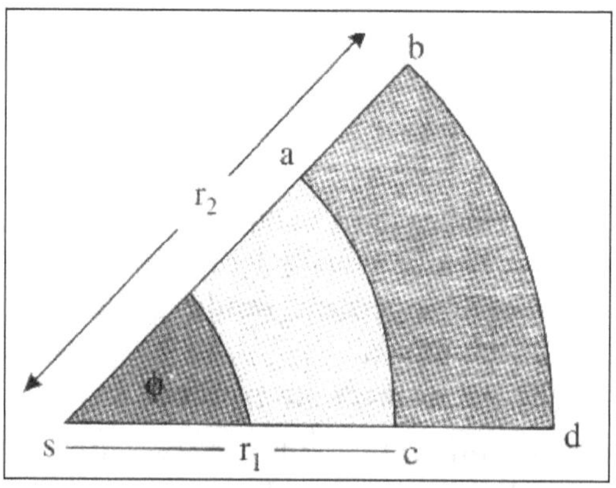

Topographic correction by compartmentalizing an area. For gravitational effect of compartment, abcd with respect to station S. r_1 and r_2 are radii of inner and outer bounding compartments respectively in centimetres.

The difference still remaining (Δg), positive or negative, is called as gravityanomaly, after applying all the corrections and getting '$g_{o/c}$' (observed and corrected), reflects conditions below ground surface in terms of rock type, size, shape and distance from the observer,

$$\Delta g = g_{O/C} - g_T = +\text{ve anomaly OR } \Delta g = g_{O/C} - gT = -\text{ve anomaly}$$

Positive anomaly is an indication of dense material underneath while negative anomaly indicates material of lesser density. The survey is conducted along a linear tract or along the grid lines which can be interpreted in terms of rock types and geological structures. The values observed are usually very small, denoted as milligal or gravitational unit (1gu = 10^{-6} m/sec² = 0.1 milligal = 1/ 10,000 gal) and can be plotted in form of contours termed as isogals, to get the regional perspective of the area.

Gravity surveys have been found to be very successful in delineating the fault zones, buried ridges, buried channels, salt domes, igneous intrusions, cavities etc. Gravity surveys are also carried out for very large regions to work out overall geological structure of a region or even of a country. For this airborne gravity surveys are conducted by installing strong gravimeter in aircrafts and equipotential lines based on observed gravity are drawn in form of contour lines. Some conceptual models of observed gravity anomaly and its interpretation is given in figures.

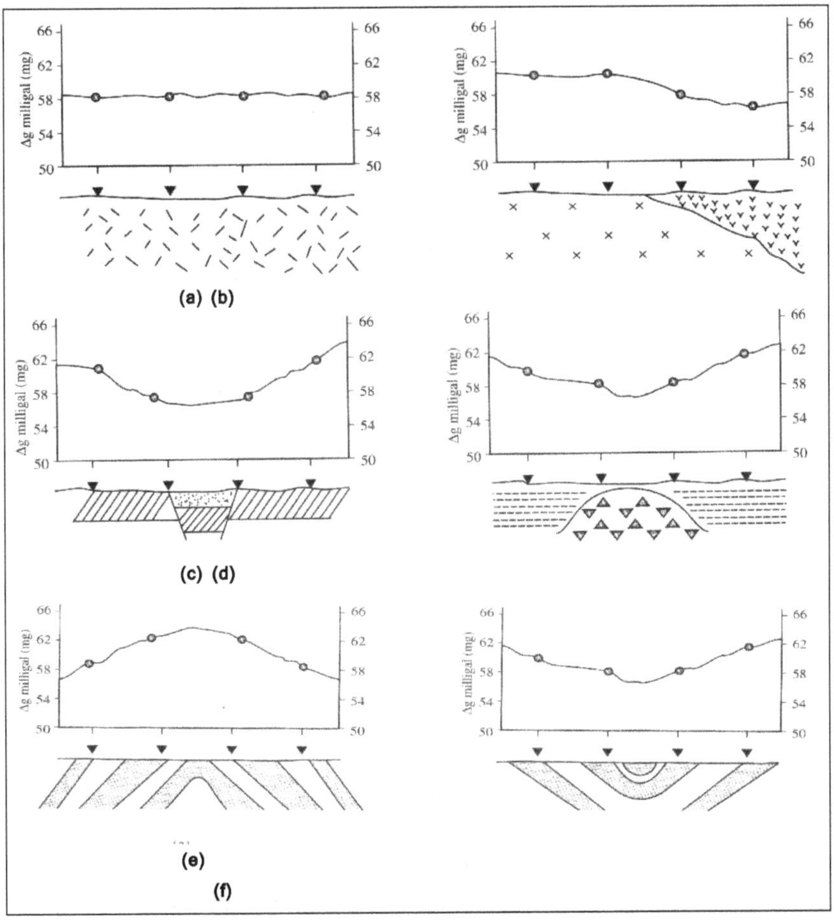

(a) (b)

(c) (d)

(e)

(f)

Figure shows that Standard gravity profiles for different sets of subsurface geological conditions. (a) Uniform gravity profile due to homogenous and monotonous rock body. (b) Sudden change in gravity profile due to sudden variation in density of rocks across a contact, fault, unconformity.(c) Sagging gravity profile due to presence of a fault zone. (d) Sagging gravity profile due to presence of a low density salt dome. (e) Up warping gravity profile due to presence of anticlinal fold and (f) Down warping gravity profile due to presence of synclinal fold.

Magnetic Method

Earth is a huge magnet, it owes its magnetism mainly due to the interaction of solid inner and liquid outer core. A substance lying in the earth's magnetic field will be susceptible (k) to magnetization which is a ratio between strength of the induced magnetization (I) and strength of the magnetic field (H) as:

$$\text{Susceptibility } k = I/H$$

Rocks have minerals and minerals have magnetic susceptibility such as haematite, 0.2 – 0.6; magnetite, 0.3 – 0.8; ilmenite, 0.115 – 0.135; siderite 0.1 – 0.4; pyrrhotite, 0.110 – 0.125 etc. Iron ores, rocks with high magnetic minerals i.e. igneous rocks which are mafic or ultramafic and objects made of iron can be easily detected if magnetic anomaly is high. For using magnetism for geophysical exploration vertical component of magnetic intensity is measured as gamma (γ), a small unit of gauss or as nanoteslar(nT).

$$1\text{ nT} = 1\gamma = 10^{-5}\text{ gauss} = 1/50{,}000 \text{ of earth's magnetic field}$$

The magnetometer used in exploration is called as Variometer which is tailor-made separately for measuring horizontal and vertical component of magnetic anomaly.

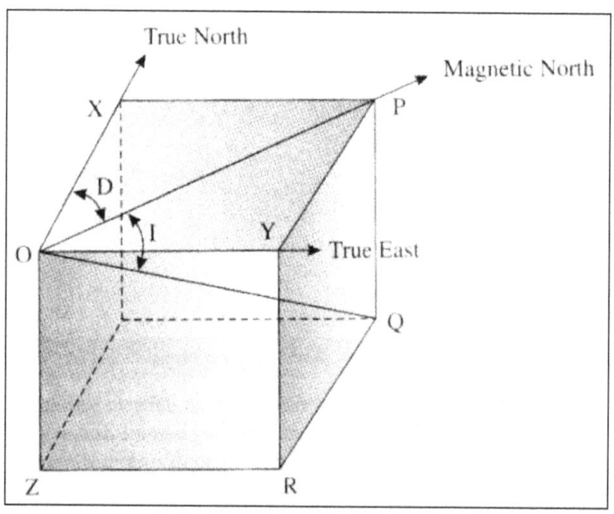

Components of earth magnetic field applied to northern hemisphere. X points north and Y points east. Q is the total intensity of the magnetic force whose horizontal component is P and vertical component Z. The magnetic dip I is measured in the plane containing P and Q while magnetic declination D is measured in horizontal plane containing PXOY.

The observed values are algebraically subtracted from theoretical values which are maintained by the Survey of India which conducts geodetic survey from time to time to be used in marine navigation. The measurements are corrected for certain systematic variations for example temperature, diurnal changes and secular horizontal declination. Air bornemagnetic exploration methods have been used to detect intrusive igneous rock bodies especially dikes by comparing the known susceptibility of rocks for example granite, 0.0027; gabbro, 0.0073; basalt, 0.0143 etc. For exploring man made features such as buried or abandoned shafts, pipelines this method is found to be very reliable.

Seismic Methods

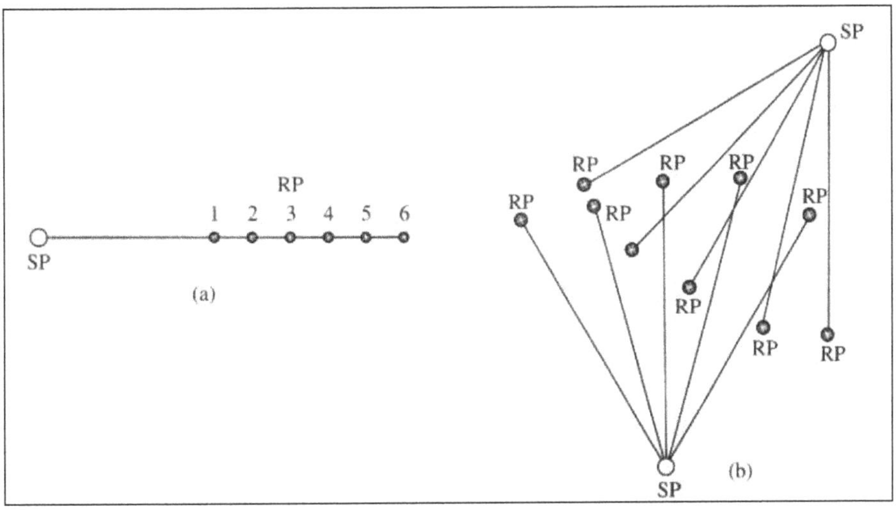

Standard seismic surveying. (a) Profile shooting and (b) Fan shooting. SP – Shot point, RP – Receiving points.

As we know that seismic waves originating from earthquake events have been used as 'X' rays for exploring the earth's deep interior. The small doses of seismic waves produced by hammer drops, explosions etc. are used for exploring shallow parts of the earth crust for mineral, oil – gas and civil engineering purposes. Seismic waves when travel through different layers of the earth they get reflected and refracted from the strata boundaries and return back to the ground surface where their arrival are recorded with the help of geophones. Geophones are placed at equi distance in linear or radial manner depending upon nature of exploration i.e. profile shooting for linear exploration and fan shooting for covering large area for survey and are connected by wires to amplifiers and recording camera and a device for noting time every hundredth of a second, usually placed in a designed truck. The distance up to which geophones are to be installed is usually ten times the depth to be explored. Seismic reflection methods are mostly used for very deep explorations usually for petroleum. For exploring shallow depths or for civil engineering purposes seismic refraction method is used. Seismic waves travel through different mediums with different velocities. It has been found that as we go down from one soil horizon to another they become denser and denser with depth due to over burden. The soil then may be taken over by some rock medium which may change to another rock with depth. Hence, when the wave come travelling through different mediums they record higher and higher velocities, which in turn are recorded in terms of different rock types from the standard chart. The velocities in weathered or heavily fractured rocks will be less while rock saturated with water will give higher seismic wave velocity as compared to non saturated rocks.

Seismic waves have been used for exploration of the subsurface for the minerals, ores and hydrocarbons. For exploring small areas and depths, these waves are generated artificially and are recorded by network of geophones in terms of their arrival time, arrival time gap and arrival locations of direct and indirect waves to know the subsurface distribution of earth materials.

Table: Average seismic wave velocities in different soil and rock mediums.

Soil/Rock	Seismic Wave Velocity (km/sec)	Soil / Rock	Seismic Wave Velocity (km/sec)
Loose Silt	0.15 – 0.30	Limeston	2.50 – 6.50
Dry Sand	0.15 – 0.45	Phyllite	3.00 – 5.80
Wet Sand	0.50 – 0.70	Schist	3.50 – 5.70
Clay	1.50 – 2.50	Gneiss	4.00 – 6.50
Gravel	0.50 – 1.50	Granite	4.50 – 6.00
Sandstone	1.50 – 4.00	Basalt	5.50 – 6.50
Shale	2.00 – 4.50	Gabbro	6.50 – 7.00

Scientists have used a stronger source of seismic waves to fathom the bigger and deeper areas of the earth. The principle of exploring the subsurface can be understood with the help of figure. When seismic waves are generated through artificial means, waves travel along surface with velocity V_1 and can be recorded by different seismographs as the "direct" waves. These waves reach these recording devices (I, II, III), closer to the shot site before the "indirect" body waves with velocity V_2, which are coming through the subsurface lower layers hence travelling longer path requiring more time. But, at device IV both the "direct" and the "indirect" waves reach at the same time because the indirect wave covering longer path has compensated the time loss by achieving higher velocity due to higher density of the subsurface denser material. After the device IV all the devices will first record arrival of the "indirect" waves, followed by the "direct" waves (V, VI, VII etc.) which are much away from the site of the shot. The distance between the shot site and the device IV is called as the "Critical" distance "X" and by using the following formula the depth of first discontinuity surface can be worked out.

$$H = \frac{X}{2}\sqrt{(V_2 - V_1) \div (V_2 + V_1)}$$

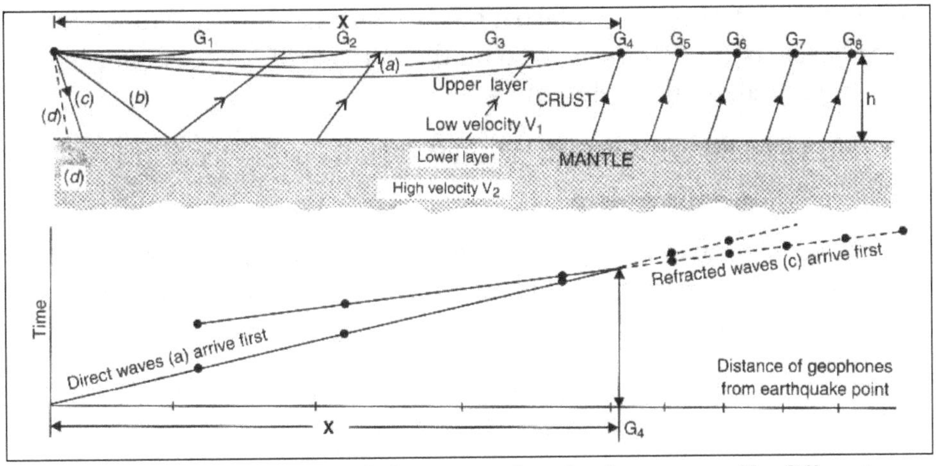

Schematics showingwave path from an earthquake site, measured by different geophones (G_1 to G_8) with corresponding time distance graph for arrival of direct and indirect waves. Also see the "Critical Distance – X".

Electrical Methods

The electrical methods of geophysical exploration measure either the electricity generating capacity termed as Self Potential method or resistance to flow of induced electrical currents of earth's natural material termed as electrical resistivity method. The self-potential method is normally used to explore the presence of iron and sulphide ores as these materials are capable of generating electricity, though very feeble but can be detected by using sophisticated instruments. From civil engineering point of view electrical resistivity method is more pertinent especially for detecting thickness of overburden, subsurface distribution of rocks, groundwater resource, leakages from pipe, canal, reservoir etc. The rocks are as such resistant to conduction of electricity but the presence of moisture in rocks or high percentage of iron bearing minerals in a rock will reduce the resistance. It is based on the premise that the free water which moves freely from pore to pore constitutes effective porosity. As an order of magnitude, the effective porosity can be for instance 70 to 80% of the total porosity in gravels and granular sand, 50 to 60% in fissured rock and 30 to 20% in fine sand and silt. This free water due to various dissolved salts is ionically conductive and enables electric currents to flow into the ground. Consequently, measuring the groundresistivitycan give an idea of the presence of water.

Conductivity (micro Siemens/cm) = 10^4/ Resistivity (ohm.m)

Conductivity (micro Siemens/cm) = 1.4 × Total Dissolved Salts (mg/l)

A standard for drinkable water is 10 ohm.m, or 1000 micro seisms/cm, or 0.7 g/l. Following basic principles are:

- A hard rock without pores or fracture and a dry sand without water or clay are highly resistive to flow of electrical currents in order of several tens of thousands ohm.m.

- A porous or fractured rock bearing free water has a resistivity which depends on the resistivity of the water and on the porosity of the rock in order of several tens to several hundred ohm.m.
- An impermeable clay layer, which has bound water, has a low resistivity few units to several tens ohm.m.
- Ametallic mineral or ore bodies (iron, manganese, copper sulphides) have very low resistivity due to their high electrical conductivity mostly lower or much lower than 1 ohm.m.

The resistivity of a porous non-clayey material can be estimated by the following Archie law formula:

$$\text{Rock resistivity} = a \times (\text{water resistivity}) / (\text{porosity})^n$$

where "a" and "n" are constants which depend on the nature of the rock. In a very rough approximation, "a" can be taken equal to 1 and "n" to 2. For example, 10 ohm.m water and 20% porosity give a rock resistivity of the order of 250 ohm.

For measuring the ground resistivity, an electric current has to be transmitted with two electrodes, while the potential created on the surface by the circulation of this current into the ground is measured with two other electrodes. By increasing the distance between the transmitting and the receiving electrodes progressively, permits to explore greater depth, termed as Depth Profiling or Sounding Array, also to determine aquifer depth and its thickness. Setting the four electrodes together at fixed distances along a traverse line helps in to detect lateral change of resistivity, termed as Horizontal Profiling or Profiling Array, mainly to detect fault or fracture zones. The shows of potable groundwater with resistivity of 10 to 100 ohm.m range will give aquifer resistivity in the range of 50 and 2000 ohm.

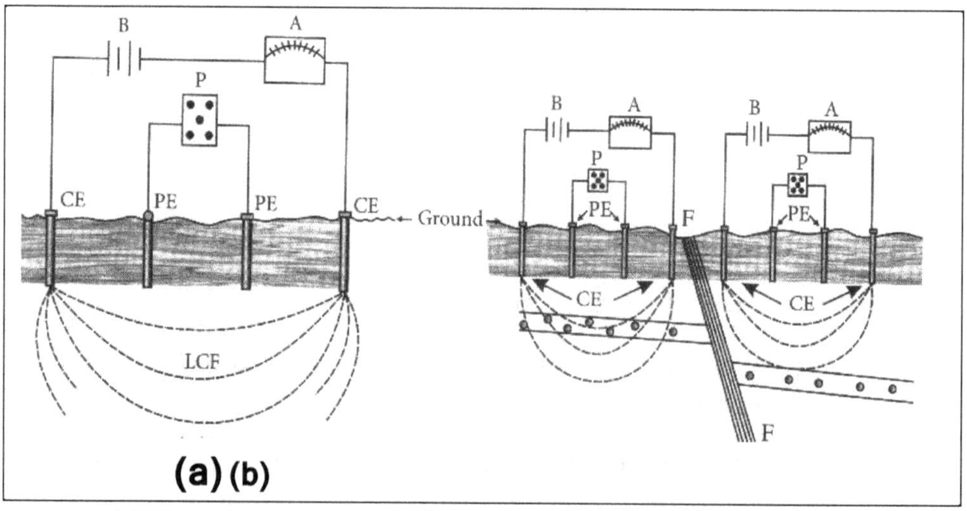

(a) Wenner's arrangement for (a) Sounding array and (b) Profiling array.

The equipments include a generator or high voltage battery as source of current, ammeter and potentiometer to measure current and potential difference four stainless steel, copper or metallic spikes for current and potential electrodes with a length of 80 cm and diameter of 2 cm and insulated cables on reels. The standard resistance of some important rocks are given in table.

Table: Standard resistivity values of some important rocks.

Rock	Resistivity (ohm-m)	Rock	Resistivity (ohm-m)	Rock	Resistivity (ohm-m)
Marble	10^4–10^8	Granite	$10^3 - 10^6$	Limestone	$10^2 - 10^4$
Gneiss	10^3–10^7	Gabbro	$10^3 - 10^5$	Sandstone	$10 - 10^3$
Slate	10^2–10^7	Basalt	$10^3 - 10^5$	Shale	$1 - 10^2$

Specific geophysical techniques are being applied successfully in most of the major projects to unravel the unseen portion of the subsurface. Once a two or three dimensional model is established one can visualize the correct picture of distribution of soil, rock head, bedrock profile, rock contacts, fault, heavily fractured zone, presence of cavities, water saturated zone, fault etc. If need be, then based on geophysical data some points can be marked for exploratory drilling to cross check the results.

The site exploration should be suited to the requirement of the project and the techniques employed should be in accordance with the size and importance of the project. If the project is small in terms of area covered and limited expenditure then data collected during reconnaissance stage may be enough but if the project is big covering large area and has ramifications on subsurface structure of the ground then a full fledge site investigation may be required.

References

- What-is-Hydrogeology, Education: sgm.gob.mx, Retrieved 26, February 2020
- Properties-of-aquifers: viu.ca, Retrieved 01, June 2020
- Engineering-geology: geologypage.com, Retrieved 30, March 2020

Chapter 4

Structural Geology

Structural geology is the study of rock deformation on a small scale as well as a large scale. The small scale study of the features of rocks can be done by mounting the ground rocks on glass slide and examining them with a polarizing microscope. This chapter has been carefully written to provide an easy understanding of the varied facets of structural geology.

Structural geology is the branch of geology that deals with the recognition, representation, and genetic interpretation of rock structures. It also involves the study of the forces which give rise to these structures. The term 'Structural' is derived from Latin word '*Struere*' which means 'to build'.

Structural geology is the study of the architecture of rocks insofar as it has resulted from deformation (Billings, 1990).

Structural geology deals with the geometry, distribution and formation of structures (Fossen, 2010).

Geologic structures or rock structures incorporate symmetry and geometric configuration of rocks present in the Earth's crust on all scales. Geologic structures result from the deformation caused by the tectonic forces present in Earth, i.e., they are endogenic. The term tectonics is derived from Greek word '*Tektos*' meaning 'builder'. Tectonics is the study of the forces and motion that result in rock deformation and structure.

Relationship between Structural Geology and Tectonics

Let us get acquainted with the relation between structural geology and tectonics. Structural geology is mainly concerned with the rock geometry whereas, tectonics deals with the forces and movements responsible for the generation of the rock or geologic structures. Therefore both structural geology and tectonics are responsible for building up the Earth's lithosphere. We can say that tectonics is quite closely connected to the underlying processes that cause geologic structures to form. These structures provide information about the forces acting within the Earth. The objective of structural geology is to determine and explain the architecture of rocks as observed in the field. The field observations are supported by laboratory investigations to attain this objective. Geologic structures range in size from microscopic scale - to hundreds of kilometres.

Use and Importance

The study of geologic structures has been of prime importance in economic geology, both petroleum geology and mining geology. Folded and faulted rock strata commonly form traps that accumulate and concentrate fluids such as petroleum and natural gas. Similarly, faulted and structurally complex areas are notable as permeable zones for hydrothermal fluids, resulting in concentrated areas of base and precious metal ore deposits. Veins of minerals containing various metals commonly occupy faults and fractures in structurally complex areas. These structurally fractured and faulted zones often occur in association with intrusive igneous rocks. They often also occur around geologic reef complexes and collapse features such as ancient sinkholes. Deposits of gold, silver, copper, lead, zinc, and other metals, are commonly located in structurally complex areas.

Structural geology is a critical part of engineering geology, which is concerned with the physical and mechanical properties of natural rocks. Structural fabrics and defects such as faults, folds, foliations and joints are internal weaknesses of rocks which may affect the stability of human engineered structures such as dams, road cuts, open pit mines and underground mines or road tunnels.

Geotechnical risk, including earthquake risk can only be investigated by inspecting a combination of structural geology and geomorphology. In addition, areas of karst landscapes which reside atop underground caverns, potential sinkholes, or other collapse features are of particular importance for these scientists. In addition, areas of steep slopes are potential collapse or landslide hazards.

Environmental geologists and hydrogeologists need to apply the tenets of structural geology to understand how geologic sites impact (or are impacted by) groundwater flow and penetration. For instance, a hydrogeologist may need to determine if seepage of toxic substances from waste dumps is occurring in a residential area or if salty water is seeping into an aquifer.

Plate tectonics is a theory developed during the 1960s which describes the movement of continents by way of the separation and collision of crustal plates. It is in a sense structural geology on a planet scale, and is used throughout structural geology as a framework to analyze and understand global, regional, and local scale features.

Concepts in Structural Geology

Let us get familiarised with some of the basic concepts of structural geology before we study about various geologic or rock structures, like folds, faults, *etc*.

The study of structural geology is mainly accomplished with the study of geological structures which are developed due to deformation. They are also known as deformational structures or secondary structures. Primary structures form as the result of the processes connected with the deposition of sediments. They are also known as depositional structures, such as bedding plane.

Structural geologists play a significant role in the identification of the geologic structures, their geometry and orientation, the time and sequence of their deposition. They also assess the physical conditions responsible for the development of these structures.

Let us read about the common terminologies used in structural geology.

Outcrop

They are exposure of rocks on the surface of Earth (Figure). If the rocks are not exposed on the surface of Earth, they are referred to as 'incrop'. The outcrops are generally visible along valley walls, river/*nala* section, mountains, road/ railway cuttings and tunnels. Hence an outcrop denotes area on surface of the Earth, over which a rock mass is exposed. The line of intersection of the limiting surface of rock mass with surface of ground marks its' limit. The study of geological structures is generally carried by the study of outcrops.

Outcrop of sandstone, note the horizontal bedding planes

An outcrop or rocky outcrop is a visible exposure of bedrock or ancient superficial deposits on the surface of the Earth.

Granite outcrops at Silesian Stones Mountain in southwestern Poland.

View of a bedrock outcrop near San Carlos Water, Falkland Islands

Serrote Branco's outcrop. Caicó, Brazil.

Features

Outcrops do not cover the majority of the Earth's land surface because in most places the bedrock or superficial deposits are covered by a mantle of soil and vegetation and cannot be seen or examined closely. However, in places where the overlying cover is removed through erosion or tectonic uplift, the rock may be exposed, or *crop out*. Such exposure will happen most frequently in areas where erosion is rapid and exceeds the weathering rate such as on steep hillsides, mountain ridges and tops, river banks, and tectonically active areas. In Finland, glacial erosion during the last glacial maximum (ca. 11000 BC), followed by scouring by sea waves, followed by isostatic uplift has produced a large number of smooth coastal and littoral outcrops.

A typical shore outcrop scoured by ancient glaciers in Espoo, Finland.

Bedrock and superficial deposits may also be exposed at the Earth's surface due to human excavations such as quarrying and building of transport routes.

Study

Outcrops allow direct observation and sampling of the bedrock *in situ* for geologic analysis and creating geologic maps. In situ measurements are critical for proper analysis of geological history and outcrops are therefore extremely important for understanding the geologic time scale of earth history. Some of the types of information that cannot be obtained except from bedrock outcrops or by precise drilling and coring operations, are structural geology features orientations (e.g. bedding planes, fold axes, foliation), depositional features orientations (e.g. paleo-current directions, grading, facies changes), paleomagnetic orientations. Outcrops are also very important for understanding fossil assemblages, and paleo-environment, and evolution as they provide a record of relative changes within geologic strata.

Accurate description, mapping, and sampling for laboratory analysis of outcrops made possible all of the geologic sciences and the development of fundamental geologic laws such as the law of superposition, the principle of original horizontality, principle of lateral continuity, and the principle of faunal succession.

Examples

On Ordnance Survey maps in Great Britain, cliffs are distinguished from outcrops: cliffs have a continuous line along the top edge with lines protruding down; outcrops have a continuous line around each area of bare rock. An outcrop example in California is the Vasquez Rocks, familiar from location shooting use in many films, composed of uplifted sandstone. Yana is another example of outcrops, located in Uttara Kannada district in Karnataka, India.

Layering

It is developed in rocks due to the deposition of sediments, rock materials or minerals. The deposition of sediments one over another in a basin results in layered or sedimentary rocks. Bed is defined as an individual layer or strata of a sedimentary rock (Figure). Each bed is separated by the adjacent bed by a plane called bedding plane. If the individual layers are less than one cm thick they are called lamination. The process of deposition of sediments in a layer by layer fashion is known as stratification.

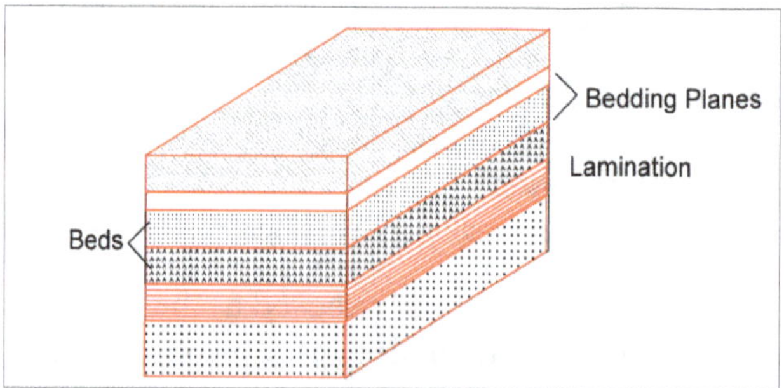

Stratification with numerous beds, bedding planes and lamination.

The bedding planes are found in sedimentary rocks. Layering is developed in volcanic rocks (igneous rocks) (Figure) because of flow of lava. It is also referred as primary foliation because these are developed simultaneously with formation of the rock. The term secondary foliation is used for layers found in metamorphic rocks (Figure). The layers in metamorphic rocks develop as a result of the development of new minerals and reorientation of mineral particles from pre-existing rocks during the process of metamorphism. Slaty cleavage, schistosity, gneissosity, *etc.* fall in the category of secondary foliation.

(a)

(b)

(a) Primary foliation developed in igneous rocks, and (b) Secondary foliation developed in metamorphic rocks.

Stratum

In geology and related fields, a stratum (plural: strata) is a layer of sedimentary rock or soil, or igneous rock that were formed at the Earth's surface, with internally consistent characteristics that distinguish it from other layers. The "stratum" is the fundamental unit in a stratigraphic column and forms the basis of the study of stratigraphy.

Strata in Salta (Argentina).

Goldenville strata in quarry in Bedford, Canada. These are Middle Cambrian marine sediments. This formation covers over half of Nova Scotia and is recorded as being 29,000 feet thick in some areas.

Characteristics

Each layer is generally one of a number of parallel layers that lie one upon another, laid down by natural processes. They may extend over hundreds of thousands of square kilometers of the Earth's surface. Strata are typically seen as bands of different colored or differently structured material exposed in cliffs, road cuts, quarries, and river banks. Individual bands may vary in thickness from a few millimeters to a kilometer or more. Each band represents a specific mode of deposition: river silt, beach sand, coal swamp, sand dune, lava bed, etc.

The Permian through Jurassic strata in the Colorado Plateau area of southeastern Utah demonstrates the principles of stratigraphy.

These strata make up much of the famous prominent rock formations in widely spaced protected areas such as Capitol Reef National Park and Canyonlands National Park.

Naming

Geologists study rock strata and categorize them by the material of beds. Each distinct layer is typically assigned to the name of sheet, usually based on a town, river, mountain, or region where the formation is exposed and available for study. For example, the Burgess Shale is a thick exposure of dark, occasionally fossiliferous, shale exposed high in the Canadian Rockies near Burgess Pass. Slight distinctions in material in a formation may be described as "members" (or sometimes "beds"). Formations are collected into "groups" while groups may be collected into "supergroups".

Beds

Beds are the layers of sedimentary rocks that are distinctly different from overlying and underlying subsequent beds of different sedimentary rocks. Layers of beds are called stratigraphy or strata. They are formed from sedimentary rocks being deposited on the Earth's solid surface over a long periods of time. The stratigraphy are layered in the same order that they were deposited, allowing a differentiation of which beds are younger and which ones are older (the Law of Superposition). The structure of a bed is determined by its bedding plane. Beds can be differentiated in various ways, including rock or mineral type and particle size. The term is generally applied to sedimentary strata, but may also be used for volcanic flows or ash layers.

Tilted sedimentary bedding in shales of the Cretaceous Salto del Fraile Formation, Peru.

In a quarry, a bedding is a term used for a structure occurring in granite and similar massive rocks that allows them to split in well-defined planes horizontally or parallel to the land surface. Other kinds of beds are cross beds and graded beds. Cross beds are not layered horizontally and are formed by a combination of local deposition on the inclined surfaces of ripples or dunes, and local erosion. Graded beds shows a gradual change in grain or clast sizes from one side of the bed to the other. A normal grading is when there are bigger grain sizes on the older side while an inverse grading is when there are smaller grain sizes on the older side. By knowing the type of beds, geologists can determine the relative ages of the rocks.

Bed Thickness

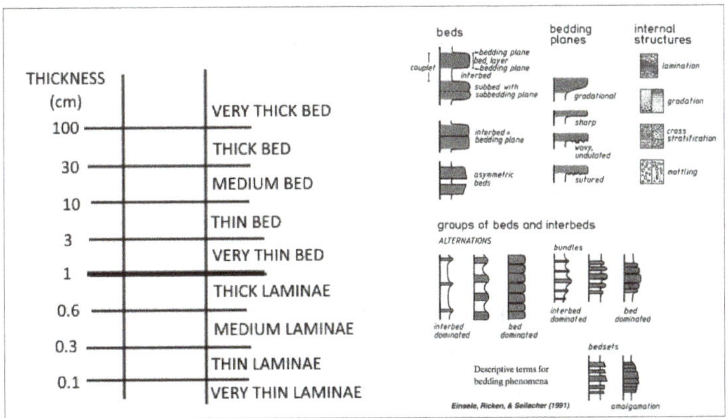

Thickness of bed and laminae sizes in centimeters.

A bed is the smallest lithostratigraphic unit, usually ranging in thickness from a centimeter to several meters and distinguishable from beds above and below it. The thickness of the bed is determined by the time period involving the deposition of the rocks.

- Very Thick Bed - 100cm

- Thick Bed - 30cm
- Medium Bed - 10cm
- Thin Bed - 3cm
- Very Thin Bed - 1cm
- Thinner than 1cm is called a Lamina

Geologic Principles

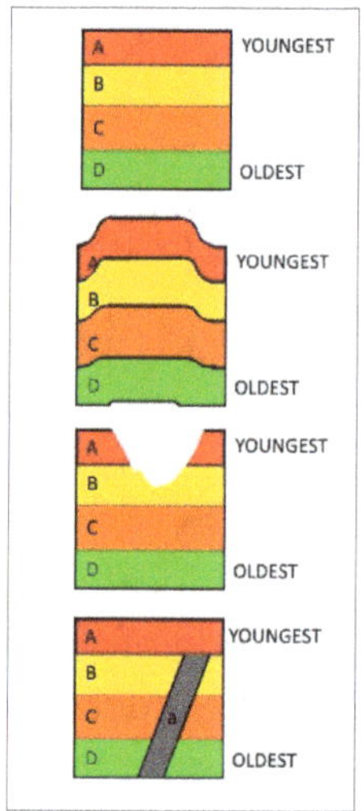

Law of Superposition, Law of Original Horizontality, Law of Lateral Continuity, Cross-Cutting Relationship.

There are geologic principles that the beds normally follow. Even though there can be cases where the principles do not apply mostly due to faults, they are true for most cases.

- Law of Superposition is the law that states that the oldest rocks are deposited first and has the younger layers deposited last, as long as the beds have not been overturned through tectonic activities. This is used to date the stratigraphy and their relative ages.
- Law of Original Horizontality states that if the beds are not horizontal, then the layers were caused to either fold or tilt through tectonic activities. They were all deposited horizontally due to gravity.
- Law of Lateral Continuity states that the bed deposits extends in all lateral directions. This means that if two places separated by erosional features have similar rocks, it could mean that they were originally continuous.
- Cross-Cutting Relationship states that a fault is younger than the rock layers that it goes through. It helps with relatively dating the rocks.

Lamination

In geology, lamination is a small-scale sequence of fine layers (*laminae*; singular: *lamina*) that occurs in sedimentary rocks. Laminae are normally smaller and less pronounced than bedding. Lamination is often regarded as planar structures one centimetre or less in thickness, whereas bedding layers are greater than one centimetre. However, structures from several millimetres to many centimetres have been described as laminae. A single sedimentary rock can have both laminae and beds.

Lamination in a piece of travertine. In this case the layering was caused by seasonal differences in sediment supply. This rock was part of the Roman aqueduct of Mons/Montauroux - Fréjus and was most probably derived from the karst area in the vicinity.

Description

Lamination consists of small differences in the type of sediment that occur throughout the rock. They are caused by cyclic changes in the supply of sediment. These changes can occur in grain size, clay percentage, microfossil content, organic material content or mineral content and often result in pronounced differences in colour between the laminae. Weathering can make the differences even more clear.

Lamination can occur as parallel structures (parallel lamination) or in different sets that make an angle with each other (cross-lamination). It can occur in many different types of sedimentary rock, from coarse sandstone to fine shales, mudstones or in evaporites.

Because lamination is a small structure, it is easily destroyed by bioturbation (the activity of burrowing organisms) shortly after deposition. Lamination therefore survives better under anoxic circumstances, or when the sedimentation rate was high and the sediment was buried before bioturbation could occur.

Origin

Lamination develops in fine grained sediment when fine grained particles settle, which can only happen in quiet water. Examples of sedimentary environments are deep marine (at the seafloor) or lacustrine (at the bottom of a lake), or mudflats, where the tide creates cyclic differences in sediment supply.

Laminae formed in glaciolacustrine environments (in glacier lakes) are a special case. They are called varves. Quaternary varves are used in stratigraphy and palaeoclimatology to reconstruct climate changes during the last few hundred thousand years.

Lamination in sandstone is often formed in a coastal environment, where wave energy causes a separation between grains of different sizes.

Foliation

Gneiss, a foliated metamorphic rock.

Foliation in geology refers to repetitive layering in metamorphic rocks. Each layer can be as thin as a sheet of paper, or over a meter in thickness. The word comes from the Latin *folium*, meaning "leaf", and refers to the sheet-like planar structure. It is caused by shearing forces (pressures pushing different sections of the rock in different directions), or differential pressure (higher pressure from one direction than in others). The layers form parallel to the direction of the shear, or perpendicular to the direction of higher pressure. Nonfoliated metamorphic rocks are typically formed in the absence of significant differential pressure or shear. Foliation is common in rocks affected by the regional metamorphic compression typical of areas of mountain belt formation (orogenic belts).

More technically, foliation is any penetrative planar fabric present in metamorphic rocks. Rocks exhibiting foliation include the standard sequence formed by the prograde metamorphism of mudrocks; slate, phyllite, schist and gneiss. The *slatey cleavage* typical of slate is due to the preferred orientation of microscopic phyllosilicate crystals. In gneiss, the foliation is more typically represented by compositional banding due to segregation of mineral phases. Foliated rock is also known as S-tectonite in sheared rock masses.

Examples include the bands in gneiss (gneissic banding), a preferred orientation of planar *large* mica flakes in schist (Schistocity), the preferred orientation of *small* mica flakes in phyllite (with its planes having a silky sheen, called *phylitic luster* – the Greek word, *phyllon*, also means "leaf"), the extremely fine grained preferred orientation of clay flakes in slate (called "slaty cleavage"), and the layers of flattened, smeared, pancake-like clasts in metaconglomerate.

Formation Mechanisms

Foliation is usually formed by the preferred orientation of minerals within a rock.

Usually this is a result of some physical force, and its effect upon the growth of minerals. The planar fabric of a foliation typically forms at right angles to the maximum principal strain direction. In sheared zones, however, planar fabric within a rock may not be directly perpendicular to the principal stress direction due to rotation, mass transport and shortening.

Foliation may be formed by realignment of micas and clays via physical rotation of the minerals within the rock. Often this foliation is associated with diagenetic metamorphism and low-grade burial metamorphism. Foliation may parallel original sedimentary bedding, but more often is oriented at some angle to it.

The growth of platy minerals, typically of the mica group, is usually a result of prograde metamorphic reactions during deformation. Often, retrograde metamorphism will not form a foliation because unroofing of a metamorphic belt is not accompanied by significant compressive stress. Thermal metamorphism in the aureole of a granite is also unlikely to result in growth of mica in a foliation, although growth of new minerals may overprint existing foliation(s).

Alignment of tabular minerals in metamorphic rocks, igneous rocks and intrusive rocks may form a foliation. Typical examples of metamorphic rocks include porphyroblastic schists where large, oblate minerals form an alignment either due to growth or rotation in the groundmass.

Igneous rocks can become foliated by alignment of cumulate crystals during convection in large magma chambers, especially ultramafic intrusions, and typically plagioclase laths. Granite may form foliation due to frictional drag on viscous magma by the wall rocks. Lavas may preserve a flow foliation, or even compressed eutaxitic texture, typically in highly viscous felsic agglomerate, welded tuff and pyroclastic surge deposits.

Metamorphic differentiation, typical of gneisses, is caused by chemical and compositional banding within the metamorphic rock mass. Usually this represents the protolith chemistry, which forms distinct mineral assemblages. However, compositional banding can be the result of nucleation processes which cause chemical and mineralogical differentiation into bands. This typically follows the same principle as mica growth, perpendicular to the principal stress. Metamorphic differentiation can be present at angles to protolith compositional banding.

Crenulation cleavage and *oblique foliation* are particular types of foliation.

Interpretation

Foliation, as it forms generally perpendicular to the direction of principal stress, records the direction of shortening. This is related to the axis of folds, which generally form an *axial-planar* foliation within their axial regions.

Measurement of the intersection between a fold's axial plane and a surface on the fold will provide the fold plunge. If a foliation does not match the observed plunge of a fold, it is likely associated with a different deformation event.

Foliation in areas of shearing, and within the plane of thrust faults, can provide information on the transport direction or sense of movement on the thrust or shear. Generally, the acute intersection angle shows the direction of transport. Foliations typically bend or curve into a shear, which provides the same information, if it is of a scale which can be observed.

Foliations, in a regional sense, will tend to curve around rigid, incompressible bodies such as granite. Thus, they are not always 'planar' in the strictest sense and may violate the rule of being perpendicular to the regional stress field, due to local influences. This is a megascopic version of what may occur around porphyroblasts. Often, fine observation of foliations on outcrop, hand specimen and on the microscopic scale complements observations on a map or regional scale.

Description

When describing a foliation it is useful to note:

- the mineralogy of the folia; this can provide information on the conditions of formation.
- the mineralogy in intrafolial areas.
- foliation spacing.
- any porphyroblasts or minerals associated with the foliation and whether they overprint it or are cut by it.
- whether it is planar, undulose, vague or well developed.
- its orientation in space, as strike and dip, or dip and dip direction.
- its relationship to other foliations, to bedding and any folding.
- measure intersection lineations.

Following such a methodology allows eventual correlations in style, metamorphic grade, and intensity throughout a region, relationship to faults, shears, structures and mineral assemblages.

Engineering Considerations

In geotechnical engineering a foliation plane may form a discontinuity that may have a large influence on the mechanical behavior (strength, deformation, etc.) of rock masses in, for example, tunnel, foundation, or slope construction.

Intrusion

Igneous bodies can be extrusive or intrusive. The intrusive rocks are those igneous rocks which penetrate or intrude in a pre-existing host rock. They might have consolidated either at great depth or at shallow depth. Many times the layering seen in the intrusive igneous rocks which can be observed in parallelism with the surrounding country rocks (also called host rock). But sometimes igneous bodies cut across the beds of the country rock and because of their late entry into the system along some joints or fractures. We will discuss various types of intrusions based on their geometry, later in this module.

Types of Intrusions

Dykes

Dykes are sheet-like bodies of igneous rock that cut across sedimentary bedding or foliations in rocks. They may be single or multiple in nature. Dykes often have a fine-grained margin where they were chilled rapidly against the wallrocks. Small dykes tend to be fine-grained while large dykes tend to be coarser grained.

Silica-rich Intrusions

- Stoped stocks

Stoping happens when a rising magma breaks off jointed blocks from the overlying country rock. The magma forces its way into the cracked roof and fragments of the wallrock sink into the magma. These fragments within the magmatic rock are known as xenoliths and can range in size from less than a millimetre up to tens of kilometres. The xenolith-rich Corralillo granodiorite of Peru is a good example, where xenoliths are concentrated near the roof of the intrusions (roof pendants) but become smaller and more rounded when digested further within the intrusion.

In the Thorr granodioritic pluton of Donegal, Ireland many of the xenoliths are large (several hundred metres) and their origin (e.g. limestone, schist) is still recognisable and can be correlated with the surrounding rock structures. This means there was little movement in the xenoliths after they were incorporated from their parent rocks.

- Ring Dykes and Bell-jar Plutons

Ring dykes form a cylinder around a subsided block of country rock, fill ing the ring-shaped fracture with magma. Excellent examples of ring dykes are seen on the island of Mull, Scotland. Large subsidence of a core of country rock can also form a circular pluton in the shape of a bell-jar.

- Centred Complexes

In composite intrusions, some are arranged in concentric rings. These often formed as rings of intrusion spread outwards from a focus. These are called centred complexes.

- Sheeted Intrusions

Some intrusions form relatively flat-lying sheets, often with an undulating surface.

- Diapiric plutons

Diapirs are circular-shaped intrusions with vertical walls that force their way into place, strongly deforming the surrounding country rocks. The diapirs probably rose as buoyant magma, forcing their way up and through the denser country rocks. Diapirs are generally unfoliated (unbanded) in their centre. They become more foliated (banded) towards the contact with the country rocks, where numerous xenoliths have become more flattened. The foliation and flattening of the xenoliths runs parallel to this contact. The adjacent surrounding country rocks are intensely deformed and stretched out parallel to the intrusion.

- Batholiths

In many parts of the world, granites extend for many hundreds of kilometres in large masses called batholiths. These are not single intrusions but usually composite intrusions of similar magmas. The individual plutons in batholiths vary in size but the largest are about 30 km across. The individual plutons can show quite different forms and ages of emplacement (sometimes separated by millions of years). At the surface, granites in batholiths often weather and erode into rounded masses called tors.

The massive batholiths represent repeated and voluminous production of magmas during a period of plate tectonic activity. An excellent example is the Berridale Batholith of the Snowy Mountains, New South Wales.

Silica-poor Intrusions

These denser magmas crystallise into several different shapes:

- Flat-lying Sheets

Flat-lying sheets (sills) range in size from less than a metre thick up to huge intrusions underlying thousands of square kilometres. The Whin Sill of northern England is up to 100 m thick and intrudes an area of over 5000 square kilometres. Although sills typically intrude along the surrounding rock layers, almost all large sills vary in thickness and transgress into higher layers when mapped over a large area. These transgressions can occur as abrupt steps. The dolerite sills of the Karoo South Africa, the Transantarctic mountains and Tasmania occur as undulating discordant sheets.

- Laccoliths

Laccoliths are lens-shaped intrusions where magmas were emplaced like a sill between sedimentary layers but then bulged up into a dome. This commonly happens in dioritic intrusions. An excellent example of a laccolith is the Prospect intrusion of Sydney, New South Wales.

- Cone Sheets

Cones are thin intrusive sheets that expand upwards and outwards in cones. Individual sheets are generally a few metres thick, but arranged so that the outer cones dip at lower angles to the inner ones so that they all converge towards a common source at depth.

- Funnel-shaped Intrusions

These have a much thicker top that plunges down to a very narrow neck. A classic example is the Skaergaard intrusion of Greenland.

- Funnel Dykes

Funnel dykes are elongated in outcrop like a dyke but with a V-shaped cross-section that narrows downwards. All known examples are very large (over 100 km in length). The Great Dyke of Zimbabwe is over 700 km long and the Jimberlana dyke of Western Australia is over 180 km long.

- Lopoliths

Lopoliths are the largest known intrusions of dense magma and form a thick saucer shape within the surrounding country rocks. A famous example is the Bushveld Complex of South Africa, which is over 550 km across and up to 8 km in thickness. Lopoliths contain many important economic deposits of nickel, copper, platinum, palladium and chromium. The Sudbury intrusion of Ontario, Canada formed in an oval-shaped depression probably caused by a large meteorite impact.

- Dyke Swarms

Dykes reach their highest numbers in dyke swarms, which may form lines or radial patterns. Radial dykes usually converge on volcanic centres or igneous intrusions. Linear dyke swarms are more extensive than radial dyke swarms but can also be concentrated around large intrusions or volcanic centres. In south-eastern Iceland, the Cenozoic lava succession is cut by thousands of aligned near-vertical basaltic dykes averaging less than a metre thick.

- Diatremes

Diatremes are steep pipe-like bodies filled with fragments of both igneous rocks and wallrocks. They form by explosion which results from the release of contained carbon dioxide gas (CO_2) and water vapour (H_2O) near the surface. Some very silica-poor magma types are economically important as the host rocks for diamonds.

Deformation

This term applied for a process in which the original rock is modified. The original rock body is usually undeformed and can be modified by folding, jointing, faulting or under the effect of gravity. Let us discuss the three types of deformation that are caused mainly by the horizontal movements of the lithospheric plates relative to one another. These forces are:

- Tensional forces stretch and pull rock masses apart,
- Compressive forces squeeze and shorten rock masses,
- Shearing forces push two parts of rock masses in opposite directions.

If a rock mass is subjected to directed force for a short period of time, it usually undergoes through three stages of deformation:

- Initially the deformation is elastic and on withdrawal of stress the body returns to its original form. The limiting stress called elastic limit.
- The deformation is plastic when the stress exceeds the elastic limit; that is the body only partially returns to its original shape even if the stress is withdrawn.
- When there is a stress continuously increases, one or more fracture develops and the body eventually fails by rupture.

Ductile and brittle are two commonly used terms by the structural geologist. Brittle deformation relates to the fracturing of the rock. Ductile deformation indicates bending, stretching, folding, thinning of rocks, and realigning of grains.

Brittle-Ductile Properties of the Lithosphere

We all know that rocks near the surface of the Earth behave in a brittle manner. Crustal rocks are composed of minerals like quartz and feldspar which have high strength, particularly at low pressure and temperature. As we go deeper in the Earth the strength of these rocks initially increases.

At a depth of about 15 km we reach a point called the brittle-ductile transition zone. Below this point rock strength decreases because fractures become closed and the temperature is higher, making the rocks behave in a ductile manner. At the base of the crust the rock type changes to peridotite which is rich in olivine. Olivine is stronger than the minerals that make up most crustal rocks, so the upper part of the mantle is again strong. But, just as in the crust, increasing temperature eventually predominates and at a depth of about 40 km the brittle-ductile transition zone in the mantle occurs. Below this point rocks behave in an increasingly ductile manner.

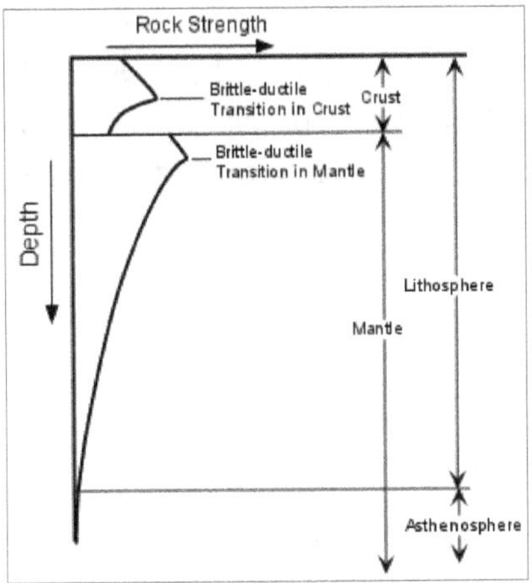

Deformation in Progress.

Only in a few cases does deformation of rocks occur at a rate that is observable on human time scales. Abrupt deformation along faults, usually associated with earthquakes occurs on a time scale of minutes or seconds. Gradual deformation along faults or in areas of uplift or subsidence can be measured over periods of months to years with sensitive measuring instruments.

Evidence of Past Deformation

Evidence of deformation that has occurred in the past is very evident in crustal rocks. For example, sedimentary strata and lava flows generally follow the law of original horizontality. Thus, when we see such strata inclined instead of horizontal, evidence of an episode of deformation.

Since many geologic features are planar in nature, we a way to uniquely define the orientation of a planar feature we first need to define two terms - strike and dip.

For an inclined plane the *strike* is the compass direction of any horizontal line on the plane. The *dip* is the angle between a horizontal plane and the inclined plane, measured perpendicular to the direction of strike.

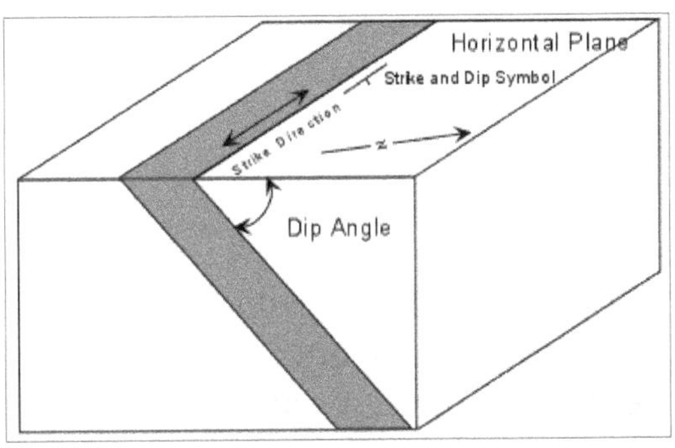

In recording strike and dip measurements on a geologic map, a symbol is used that has a long line oriented parallel to the compass direction of the strike. A short tick mark is placed in the center of the line on the side to which the inclined plane dips, and the angle of dip is recorded next to the strike and dip symbol as shown above. For beds with a 90° dip (vertical) the short line crosses the strike line, and for beds with no dip (horizontal) a circle with a cross inside is used as shown below.

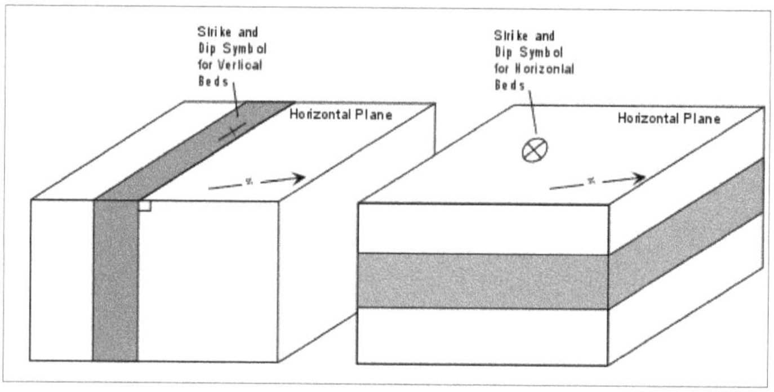

For linear structures, a similar method is used, the strike or bearing is the compass direction and angle the line makes with a horizontal surface is called the plunge angle.

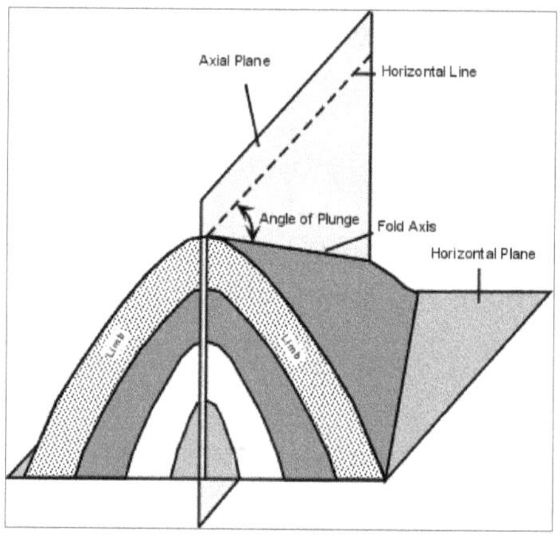

Fracture of Brittle Rocks.

As we have discussed previously, brittle rocks tend to fracture when placed under a high enough stress. Such fracturing, while it does produce irregular cracks in the rock, sometimes produces planar features that provide evidence of the stresses acting at the time of formation of the cracks. Two major types of more or less planar fractures can occur: joints and faults.

Joints

As we learned in our discussion of physical weathering, joints are fractures in rock that show no slippage or offset along the fracture. Joints are usually planar features, so their orientation can be described as a strike and dip. They form from as a result of extensional stress acting on brittle rock. Such stresses can be induced by cooling of rock (volume decreases as temperature decreases) or by relief of pressure as rock is eroded above thus removing weight.

Joints provide pathways for water and thus pathways for chemical weathering attack on rocks. If new minerals are precipitated from water flowing in the joints, this will form a vein. Many veins observed in rock are mostly either quartz or calcite, but can contain rare minerals like gold and silver. These aspects will be discussed in more detail when we talk about valuable minerals from the earth in a couple of weeks.

Because joints provide access of water to rock, rates of weathering and/or erosion are usually higher along joints and this can lead to differential erosion.

From an engineering point of view, joints are important structures to understand. Since they are zones of weakness, their presence is critical when building anything from dams to highways. For dams, the water could leak out through the joints leading to dam failure. For highways the joints may separate and cause rock falls and landslides.

Faults

Faults occur when brittle rocks fracture and there is an offset along the fracture. When the offset is small, the displacement can be easily measured, but sometimes the displacement is so large that it is difficult to measure.

Types of Faults

As we found out in our discussion of earthquakes, faults can be divided into several different types depending on the direction of relative displacement. Since faults are planar features, the concept of strike and dip also applies, and thus the strike and dip of a fault plane can be measured. One division of faults is between dip-slip faults, where the displacement is measured along the dip direction of the fault, and strike-slip faults where the displacement is horizontal, parallel to the strike of the fault. Recall the following types of faults:

- Dip Slip Faults - Dip slip faults are faults that have an inclined fault plane and along which the relative displacement or offset has occurred along the dip direction. Note that in looking at the displacement on any fault we don't know which side actually moved or if both sides moved, all we can determine is the relative sense of motion.
    - Normal Faults - are faults that result from horizontal tensional stresses in brittle rocks and where the hanging-wall block has moved down relative to the footwall block.

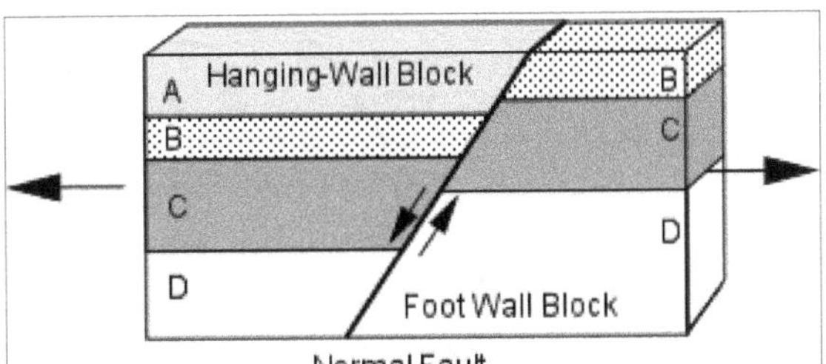

Normal Fault
Extensional Stress

Horsts & Grabens - Due to the tensional stress responsible for normal faults, they often occur in a series, with adjacent faults dipping in opposite directions. In such a case the down-dropped blocks form *grabens* and the uplifted blocks form *horsts*. In areas where tensional stress has recently affected the crust, the grabens may form *rift valleys* and the uplifted horst blocks may form linear mountain ranges. The East African Rift Valley is an example of an area where continental extension has created such a rift. The basin and range province of the western U.S. (Nevada, Utah, and Idaho) is also an area that has recently undergone crustal extension. In the basin and range, the basins are elongated grabens that now form valleys, and the ranges are uplifted horst blocks.

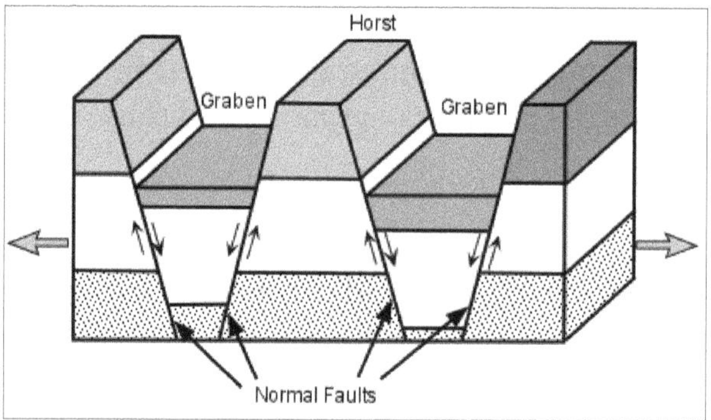

Half-Grabens - A normal fault that has a curved fault plane with the dip decreasing with depth can cause the down-dropped block to rotate. In such a case a half-graben is produced, called such because it is bounded by only one fault instead of the two that form a normal graben.

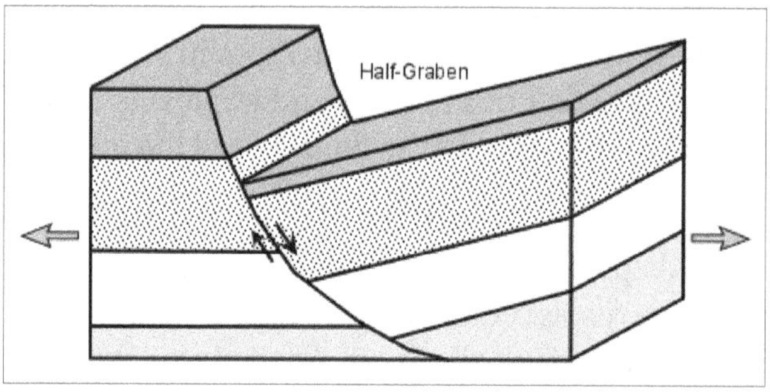

- *Reverse Faults* - are faults that result from horizontal compressional stresses in brittle rocks, where the hanging-wall block has moved up relative the footwall block.

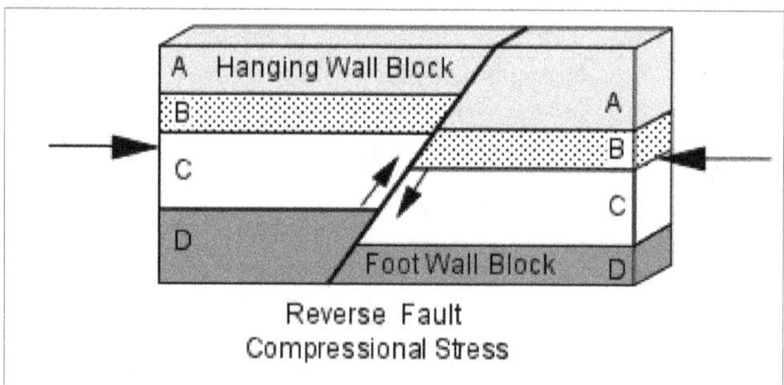

A *Thrust Fault* is a special case of a reverse fault where the dip of the fault is less than 45°. Thrust faults can have considerable displacement, measuring hundreds of kilometers, and can result in older strata overlying younger strata.

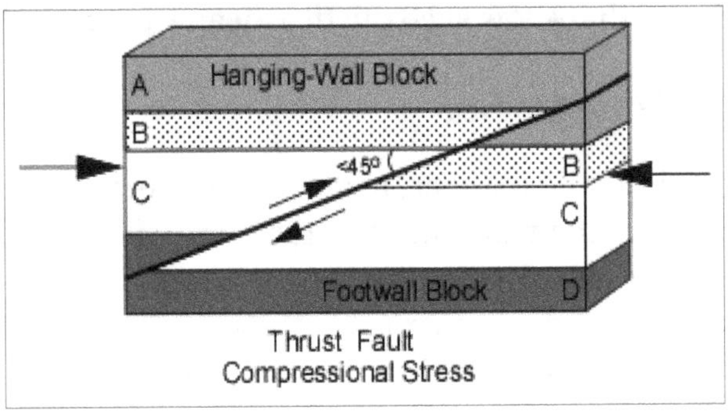

- *Strike Slip Faults* - are faults where the relative motion on the fault has taken place along a horizontal direction. Such faults result from shear stresses acting in the crust. Strike slip faults can be of two varieties, depending on the sense of displacement. To an observer standing on one side of the fault and looking across the fault, if the block on the other side has moved to the left, we say that the fault is a *left-lateral strike-slip fault*. If the block on the other side has moved to the right, we say that the fault is a *right-lateral strike-slip fault*. The famous San Andreas Fault in California is an example of a right-lateral strike-slip fault. Displacements on the San Andreas fault are estimated at over 600 km.

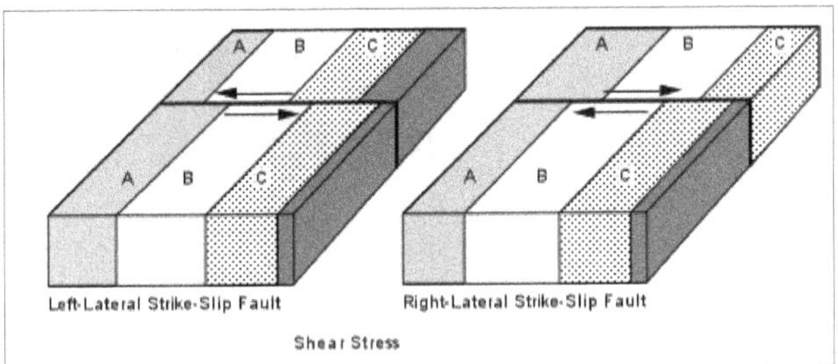

Evidence of Movement on Faults

Since movement on a fault involves rocks sliding past each other there may be left evidence of movement in the area of the fault plane.

- *Fault Breccias* are crumbled up rocks consisting of angular fragments that were formed as a result of grinding and crushing movement along a fault. When the rock is broken into clay or silt size particles as a result of slippage on the fault, it is referred to as *fault gouge.*
- *Slickensides* are scratch marks that are left on the fault plane as one block moves relative to the other. Slickensides can be used to determine the direction and sense of motion on a fault.
- *Mylonite* - Along some faults rocks are sheared or drawn out by ductile

deformation along the fault. This results in a type of localized metamorphism called dynamic metamorphism (also called cataclastic metamorphism. The resulting rock is a fine grained metamorphic rock show evidence of shear, called a mylonite. Faults that show such ductile shear are referred to as *shear zones.*

Deformation of Ductile Rocks

When rocks deform in a ductile manner, instead of fracturing to form faults or joints, they may bend or fold, and the resulting structures are called *folds.* Folds result from compressional stresses or shear stresses acting over considerable time. Because the strain rate is low and/or the temperature is high, rocks that we normally consider brittle can behave in a ductile manner resulting in such folds.

Geometry of Folds - Folds are described by their form and orientation. The sides of a fold are called *limbs.* The limbs intersect at the tightest part of the fold, called the *hinge.* A line connecting all points on the hinge is called the *fold axis.* An imaginary plane that includes the fold axis and divides the fold as symmetrically as possible is called the *axial plane* of the fold.

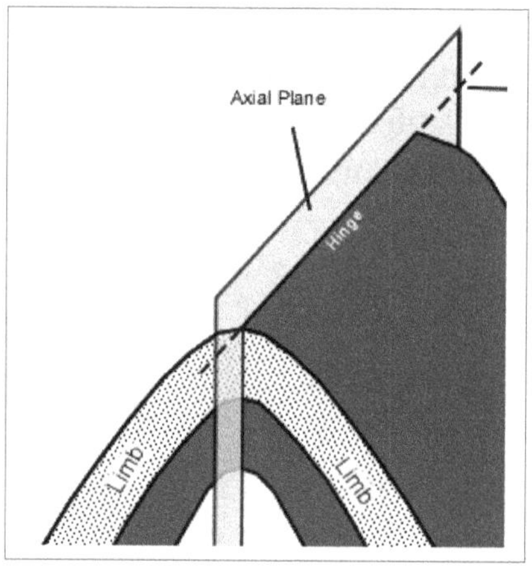

We recognize several different kinds of folds.

Monoclines are the simplest types of folds. Monoclines occur when horizontal strata are bent upward so that the two limbs of the fold are still horizontal.

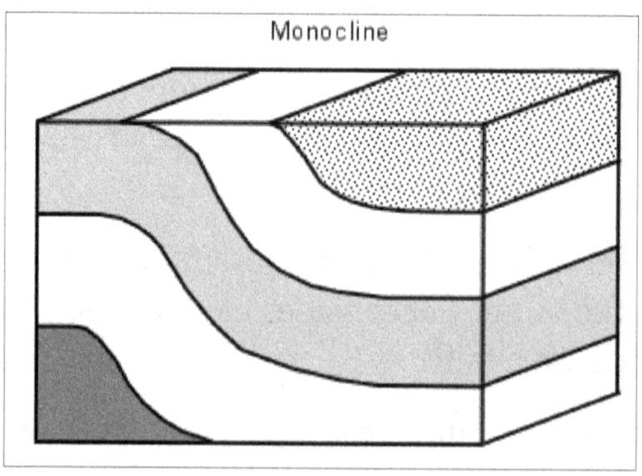

Anticlines are folds where the originally horizontal strata has been folded upward, and the two limbs of the fold dip away from the hinge of the fold.

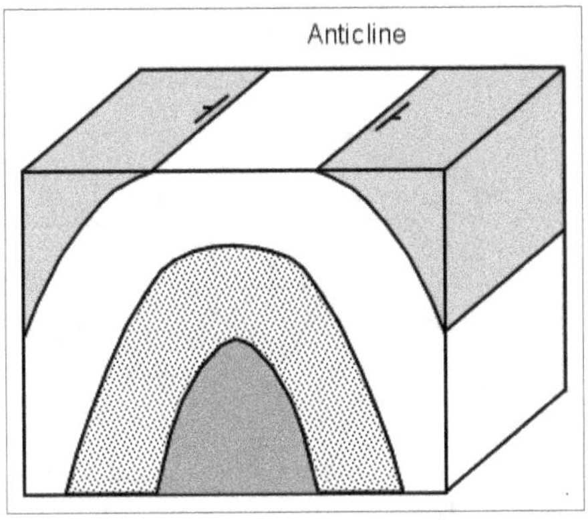

Synclines are folds where the originally horizontal strata have been folded downward, and the two limbs of the fold dip inward toward the hinge of the fold. Synclines and anticlines usually occur together such that the limb of a syncline is also the limb of an anticline.

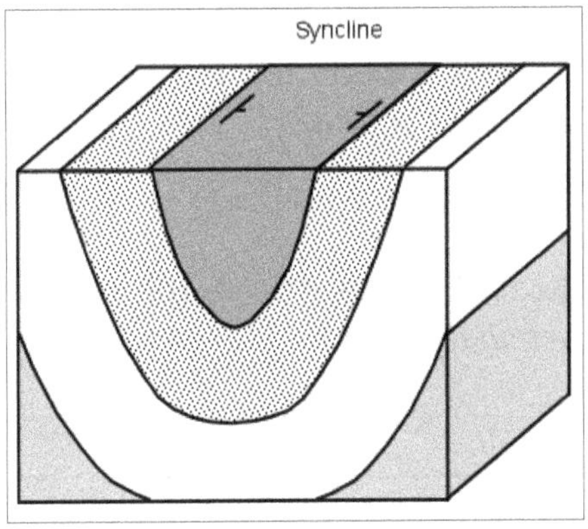

- In the diagrams above, the fold axes are horizontal, but if the fold axis is not horizontal the fold is called a *plunging fold* and the angle that the fold axis makes with a horizontal line is called the *plunge* of the fold.

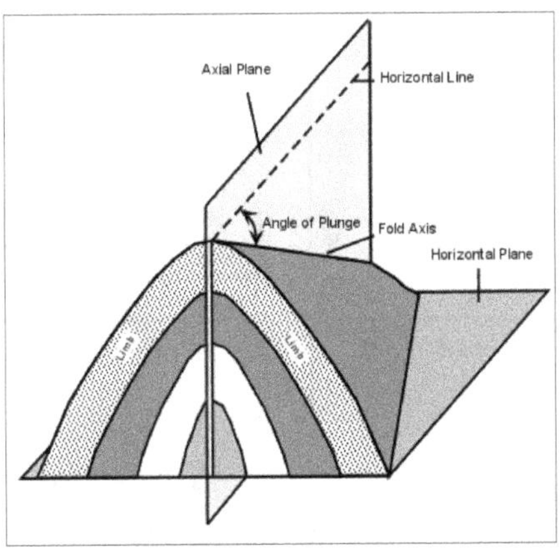

Note that if a plunging fold intersects a horizontal surface, we will see the pattern of the fold on the surface.

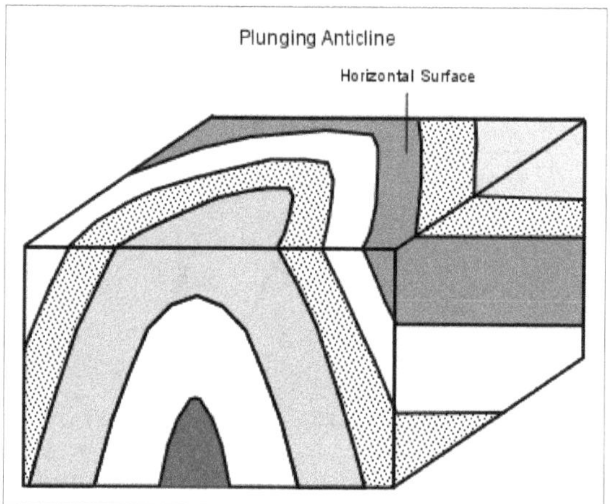

Domes and *Basins* are formed as a result of vertical crustal motion. Domes look like an overturned bowl and result from crustal upwarping. Basins look like a bowl and result from subsidence.

Folds are described by the severity of folding. an *open fold* has a large angle between limbs, a *tight fold* has a small angle between limbs.

Further classification of folds include:

- If the two limbs of the fold dip away from the axis with the same angle, the fold is said to be a *symmetrical fold.*
- If the limbs dip at different angles, the folds are said to be *asymmetrical folds.*

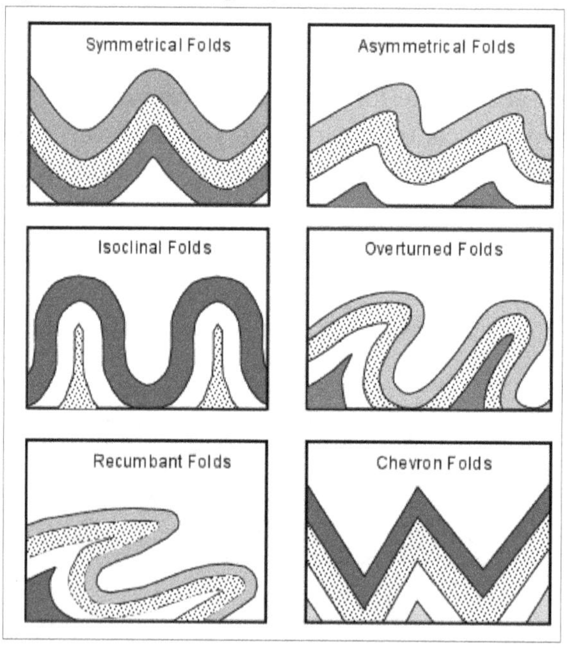

- If the compressional stresses that cause the folding are intense, the fold can close up and have limbs that are parallel to each other. Such a fold is called an *isoclinal fold* (iso means same, and cline means angle, so isoclinal means the limbs have the same angle). Note the isoclinal fold depicted in the diagram below is also a symmetrical fold.
- If the folding is so intense that the strata on one limb of the fold becomes nearly upside down, the fold is called an *overturned fold.*
- An overturned fold with an axial plane that is nearly horizontal is called a *recumbant fold.*
- A fold that has no curvature in its hinge and straight-sided limbs that form a zigzag pattern is called a *chevron fold.*

Folds and Topography

Since different rocks have different resistance to erosion and weathering, erosion of folded areas can lead to a topography that reflects the folding. Resistant strata would form ridges that have the same form as the folds, while less resistant strata will form valleys.

How Folds Form

Folds develop in two ways:

- *Flexural folds* form when layers slip as stratified rocks are bent. This results in the layers maintaining their thickness as they bend and slide over one another. These are generally formed due to compressional stresses acting from either side.
- *Flow folds* form when rocks are very ductile and flow like a fluid. Different parts of the fold are drawn out by this flow to different extents resulting in layers becoming thinner in some places and thicker in outer places. The flow results in shear stresses that smear out the layers.
- Folds can also form in relationship to faulting of other parts of the rock body. In this case the more ductile rocks bend to conform to the movement on the fault.

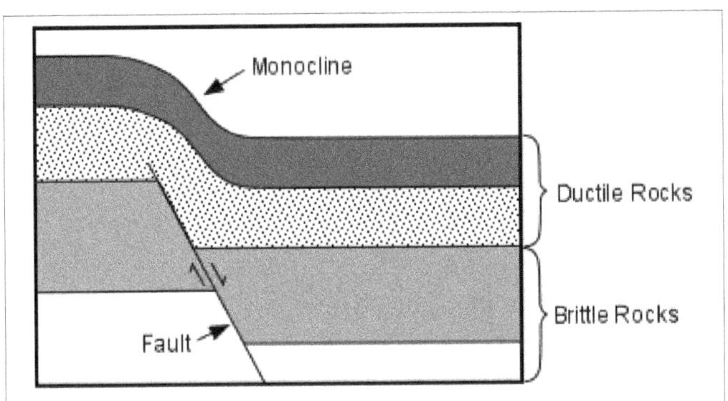

- Also since even ductile rocks can eventually fracture under high stress, rocks may fold up to a certain point then fracture to form a fault.

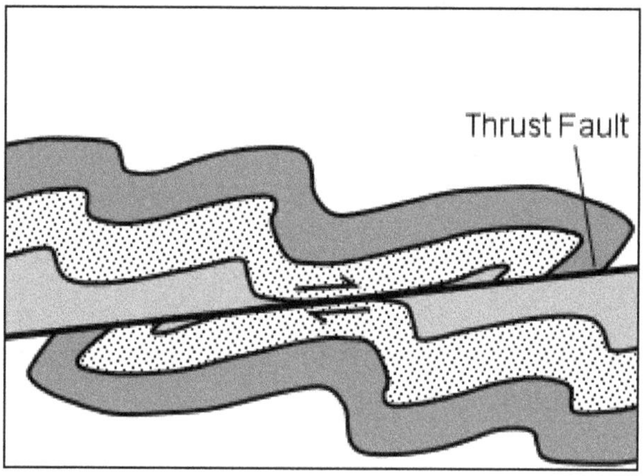

Folds and Metamorphic Foliation

As we saw in our discussion of metamorphic rocks, foliation is a planar fabric that develops in rocks subject to compressional stress during metamorphism. It may be present as flattened or elongated grains, with the flattening occurring perpendicular to the direction of compressional stress. It also results from the reorientation, recrystallization, or growth of sheet silicate minerals so that their sheets become oriented perpendicular to the compressional stress direction. Thus, we commonly see a foliation that is parallel to the axial plane of the fold.

Shearing of rock during metamorphism can also draw out grains in the direction of shear.

Mountains and Mountain Building Processes

One of the most spectacular results of deformation acting within the crust of the Earth is the formation of mountain ranges. Mountains frequently occur in elongate, linear belts. They are constructed by tectonic plate interactions in a process called orogenesis.

Mountain building (orogenesis) involves

- Structural deformation.
- Faulting.
- Folding.
- Igneous Processes.
- Metamorphism.
- Glaciation.

- Erosion.
- Sedimentation

Constructive processes, like deformation, folding, faulting, igneous processes and sedimentation build mountains up; destructive processes like erosion and glaciation, tear them back down again.

Mountains are born and have a finite life span. Young mountains are high, steep, and growing upward. Middle-aged mountains are cut by erosion. Old mountains are deeply eroded and often buried. Ancient orogenic belts are found in continental interiors, now far away from plate boundaries, but provide information on ancient tectonic processes. Since orogenic continental crust generally has a low density and thus is too buoyant to subduct, if it escapes erosion it is usually preserved.

Uplift and Isostasy

The fact that marine limestones occur at the top of Mt. Everest, indicates that deformation can cause considerable vertical movement of the crust. Such vertical movement of the crust is called *uplift*. Uplift is caused by deformation which also involves thickening of the low density crust and, because the crust "floats" on the higher density mantle, involves another process that controls the height of mountains.

The discovery of this process and its consequences involved measurements of gravity. Gravity is measured with a device known as a gravimeter. A gravimeter can measure differences in the pull of gravity to as little as 1 part in 100 million. Measurements of gravity can detect areas where there is a deficiency or excess of mass beneath the surface of the Earth. These deficiencies or excesses of mass are called *gravity anomalies*.

A positive gravity anomaly indicates that an excess of mass exits beneath the area. A negative gravity anomaly indicates that there is less mass beneath an area.

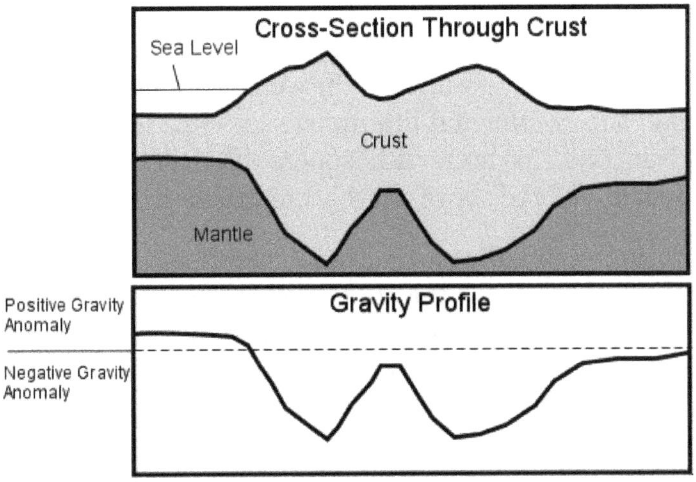

Negative anomalies exist beneath mountain ranges, and mirror the topography and crustal thickness as determined by seismic studies. Thus, the low density continents appear to be floating on higher density mantle.

The protrusions of the crust into the mantle are referred to as crustal roots. Normal crustal thickness, measured from the surface to the Moho is 35 to 40 km. But under mountain belts crustal thicknesses of 50 to 70 km are common. In general, the higher the mountains, the thicker the crust.

What causes this is the principal of *isostasy*. The principal can be demonstrated by floating various sizes of low density wood blocks in your bathtub or sink. The larger blocks will both float higher and extend to deeper levels in the water and mimic the how the continents float on the mantle.

It must be kept in mind, however that it's not just the crust that floats, it's the entire lithosphere. So, the lithospheric mantle beneath continents also extends to deeper levels and is thicker under mountain ranges than normal. Because the lithosphere is floating in the asthenosphere which is more ductile than the brittle lithosphere, the soft asthenosphere can flow to compensate for any change in thickness of the crust caused by erosion or deformation.

The Principle of isostasy states that there is a flotational balance between low density rocks and high density rocks. i.e. low density crustal rocks float on higher density mantle rocks. The height at which the low density rocks float is dependent on the thickness of the low density rocks. Continents stand high because they are composed of low density rocks (granitic composition). Ocean basins stand low, because they are composed of higher density basaltic and gabbroic rocks.

Isostasy is best illustrated by effects of glaciation. During an ice age crustal rocks that are covered with ice are depressed by the weight of the overlying ice. When the ice melts, the areas previously covered with ice undergo uplift.

Mountains only grow so long as there are forces causing the uplift. As mountains rise, they are eroded. Initially the erosion will cause the mountains to rise higher as a result of isostatic compensation. But, eventually, the weight of the mountain starts to depress the lower crust and sub-continental lithosphere to levels where they start to heat up and become more ductile. This hotter lithosphere will then begin to flow outward away from the excess weight and the above will start to collapse.

The hotter rocks could eventually partially melt, resulting in igneous intrusions as the magmas move to higher levels, or the entire hotter lower crust could begin to rise as a result of their lower density. These processes combined with erosion on the surface result in *exhumation*, which causes rocks from the deep crust to eventually become exposed at the surface.

Causes of Mountain Building

There are three primary causes of mountain building.

1. Convergence at convergent plate boundaries.

   When oceanic lithosphere subducts beneath continental lithosphere magmas generated above the subduction zone rise, intrude, and erupt to form *volcanic mountains*. The compressional stresses generated between the trench and the volcanic arc create *fold-thrust mountain belts,* and similar compression behind the arc create a fold-thrust belt resulting in mountains. Mountains along the margins of western North and South America, like the Andes and the Cascade range formed in this fashion.

   Island arcs off the coast of continents can get pushed against the continent. Because of their low density, they don't subduct, but instead get accreted to the edge of the continent. Mountain ranges along the west coast of North America formed in this fashion.

2. Continental Collisions.

   Plate tectonics can cause continental crustal blocks to collide. When this occurs the rocks between the two continental blocks become folded and faulted under compressional stresses and are pushed upward to form *fold-thrust mountains.* The Himalayan Mountains (currently the highest on Earth) are mountains of this type and were formed as a result of the Indian Plate colliding with the Eurasian plate. Similarly the Appalachian Mountains of North America and the Alps of Europe were formed by such processes.

3. Rifting

   Continental Rifting occurs where continental crust is undergoing extensional deformation. This results in thinning of the lithosphere and upwelling of the asthenosphere which results in uplift. The brittle lithosphere responds by producing normal faults where blocks of continental lithosphere are uplifted to form grabens or half grabens. The uplifted blocks are referred to a *fault-block mountains.*

   The Basin and Range province in the western United states formed in this manner, including the Sierra Nevada on its western edge and the Grand Tetons in Wyoming.

Cratons and Orogens

The continents can be divided into two kinds of structural units:

- *Cratons* form the cores of the continents. These are portions of continental crust that have attained isostatic and tectonic stability and have cooled substantially since their formation. They were formed and were deformed more than a billion years ago and are the oldest parts of the continents. The represent the deep roots of former mountains and consist of metamorphic and plutonic igneous rocks, all showing extensive evidence of deformation.

- *Orogens* are broad elongated belts of deformed rocks that are draped around the cratons. They appear to be the eroded roots of former mountain belts that formed by continent - continent collisions. Only the youngest of these orogens still form mountain ranges.

The observation that the orogens are generally younger towards the outside of any continent suggests that the continents were built by collisions of plates that added younger material to the outside edges of the continents, and is further evidence that plate tectonics has operated for at least the last 2 billion years.

Strike and Dip

A standard Brunton compass, used commonly by geologists for strike and dip measurements.

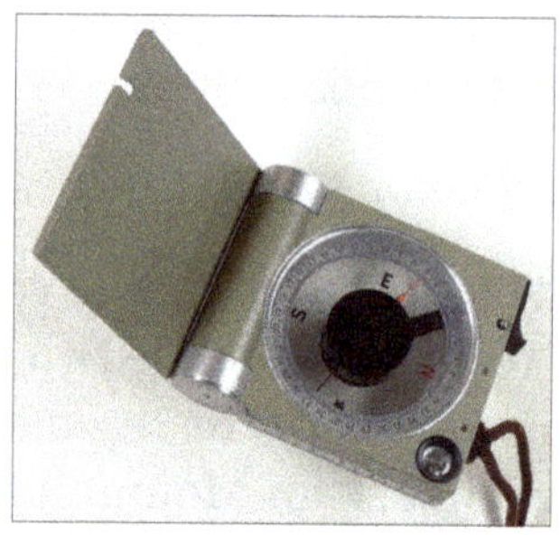

Stratum compass to measure dip and dip direction in one step.

Strike and dip refer to the orientation or *attitude* of a geologic feature. The *strike line* of a bed, fault, or other planar feature, is a line representing the intersection of that feature with a horizontal plane. On a geologic map, this is represented with a short straight line segment oriented parallel to the strike line. *Strike* (or strike angle) can be given as either a quadrant compass bearing of the strike line (N25°E for example) or in terms of east or west of true north or south, a single three digit number representing the azimuth, where the lower number is usually given (where the example of N25°E would simply be 025), or the azimuth number followed by the degree sign (example of N25°E would be 025°).

The *dip* gives the steepest angle of descent of a tilted bed or feature relative to a horizontal plane, and is given by the number (0°-90°) as well as a letter (N,S,E,W) with rough direction in which the bed is dipping downwards. One technique is to always take the strike so the dip is 90° to the right of the strike, in which case the redundant letter following the dip angle is omitted (right hand rule, or RHR). The map symbol is a short line attached and at right angles to the strike symbol pointing in the direction which the planar surface is dipping down. The angle of dip is generally included on a geologic map without the degree sign. Beds that are dipping vertically are shown with the dip symbol on both sides of the strike, and beds that are level are shown like the vertical beds, but with a circle around them. Both vertical and level beds do not have a number written with them.

Another way of representing strike and dip is by dip and dip direction. The dip direction is the azimuth of the direction the dip as projected to the horizontal (like the trend of a linear feature in trend and plunge measurements), which is 90° off the strike angle. For example, a bed dipping 30° to the South, would have an East-West strike (and would be written 090°/30° S using strike and dip), but would be written as 30/180 using the dip and dip direction method.

Strike and dip are determined in the field with a compass and clinometer or a combination of the two, such as a Brunton compass named after D.W. Brunton, a Colorado miner. Compass-clinometers which measure dip and dip direction in a single operation (as pictured) are often called "stratum" or "Klar" compasses after a German professor. Smartphone apps are also now available, that make use of the internal accelerometer to provide orientation measurements. Combined with the GPS functionality of such devices, this allows readings to be recorded and later downloaded onto a map.

Any planar feature can be described by strike and dip. This includes sedimentary bedding, faults and fractures, cuestas, igneous dikes and sills, metamorphic foliation and any other planar feature in the Earth. Linear features are measured with very similar methods, where "plunge" is the dip angle and "trend" is analogous to the dip direction value.

Apparent dip is the name of any dip measured in a vertical plane that is not perpendicular to the strike line. True dip can be calculated from apparent dip using trigonometry if you know the strike. Geologic cross sections use apparent dip when they are drawn at some angle not perpendicular to strike.

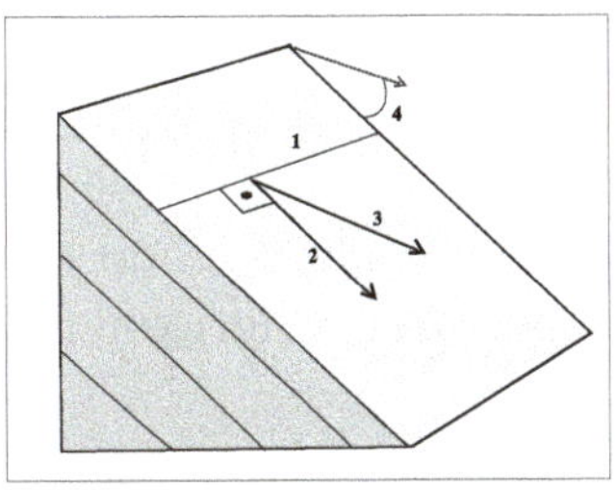

Strike and dip of the beds.
1-Strike, 2-Dip direction,
3-Apparent dip 4-Angle of dip.

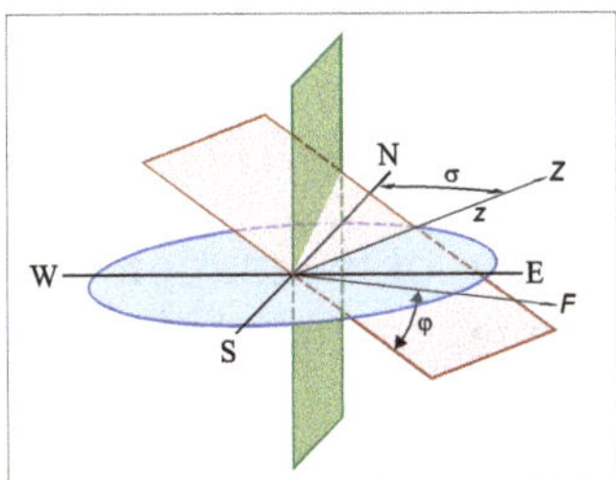

Strike and dip.

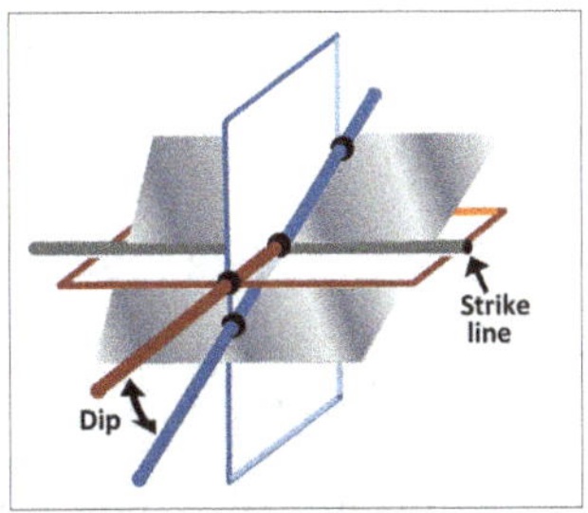

Strike line and dip of a plane describing attitude relative to a horizontal plane and a vertical plane perpendicular to the strike line.

Dip and Strike

Initially the sedimentary rocks are deposited on flat or gently inclined surfaces. Post depositional deformation is caused by the movements due to tectonic forces acting on rock body or mass which causes the tilting of the beds.

Dip: It is defined as inclination of the bedding plane with respect to horizontal. It is measured in a vertical plane lying at right angles to the strike of the bedding.

Strike: It is the geographical direction of a line produced by intersection between inclined layers and a horizontal plane (parallel to surface to Earth).

Direction of dip: It is the geographical direction, along which a bed has maximum slope.

Amount of dip: It is the angle which varies from 00 to 90°, according to the inclination of the bed.

As discussed above the disposition of the beds (Figure) can be:

- Horizontal: angle of dip=0° (no dip)
- Inclined or tilted: angle of dip varying between 0°-90°
- Vertical: angle of dip=90°

Dip is the vector quantity as it has both direction and amount. Amount of dip denotes the angle of the bed inclination with respect to the horizontal. Strike is a scalar quantity because it has only direction and no magnitude. Strike of the bed is independent of its amount of dip.

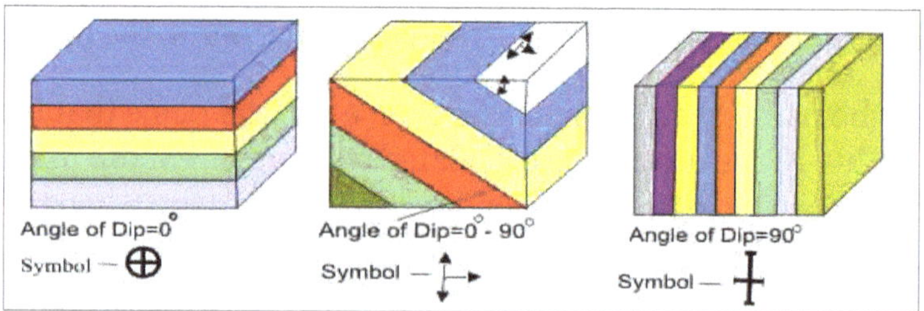

Horizontal, inclined, and vertical beds with symbolic representation on map.

True and apparent dip: Dip can be of two types (Figure). Let us read about them.

- True dip: The maximum amount of inclination or slope of bed along a line perpendicular to the strike is the maximum slope with respect to the horizon, it is called true dip.
- Apparent dip: The dip of the bed measured in any direction other than that of true dip is called apparent dip. The amount of apparent dip is always less than amount of true dip.

Relation between dip and strike: The direction of dip and strike of any inclined or tilted bed must lie at right angles to each other. Thus true dip is in the direction perpendicular to strike. While mentioning the attitude of any inclined bed, dip amount and dip and strike direction should be mentioned.

Importance of Strike and Dip: Let us discuss the importance of strike and dip in structural geology.

(a) Determination of the younger bed or formation: In geological formations the older rocks deposit at the base which is superimposed by the younger rocks. Hence, in a tilted rock sequence when we move in the direction of dip then relatively beds of younger age will be encountered and vice-versa.

(b) Classification of geological structures: Dip and strike data provides useful information in the classification of rock or geologic structures.

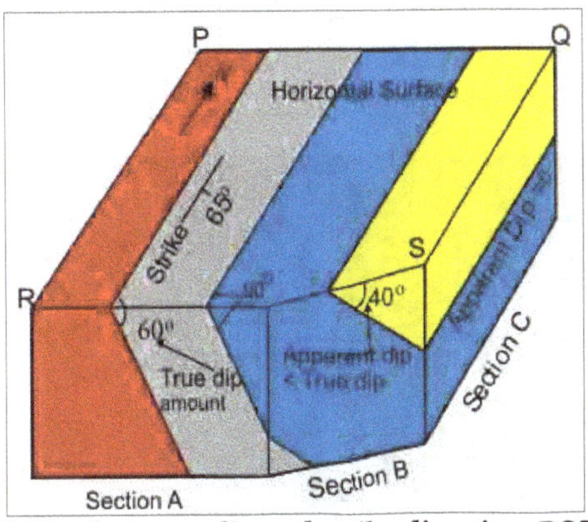

True dip, apparent dip and relation between dip and strike direction. PQRS is a horizontal plane. Beds are all dipping 600 towards east and they strike north-south.

Clinometer Compass

We have read that many rocks show planar structures in the form of bedding plane or foliation in any outcrop. They are not always horizontal, may show inclination of varying amounts. We have discussed that strike and dip describe the attitude of a rock layer in an outcrop. Let us recall that strike is the compass direction of a line formed by the intersection of a rock layer's surface with horizontal surface. Whereas the dip is measured at right angles to the strike and it is simply the amount of tilting or in other words the angle at which the rock layer is inclined from the horizontal. The amount of dip of a bedding plane varies. In order to understand the attitude and orientation of the rock structures, you are required to measure their dip and strike on the outcrops during the fieldwork. The measurement is done with the help of Clinometer Compass (Figure).

Clinometer Compass consists of a compass and clinometer. The compass has a circular dial which is anticlockwise graduated with 360 divisions in the outer most circle for reading the direction. This circle is called azimuthal circle; where 0 or 360 represents for North, 90 for East, 180 for South and 270 for West. It has a magnetic needle which rests in north- south direction of the Earth, when set free. The desired direction is read on the dial with help of N (north) marked end of this magnetic needle. One of the inner circles on the dial is also marked with directions such as E, NE, NNE, W etc. in supplement to the outermost circle to read the direction directly.

The inclination of a line with respect to horizontal is read with the help of clinometer fitted in compass (Figure). It has a frame with open space within the clinometer frame which allows seeing one of the inner circles in the dial marked with 0 to 900 graduation. The reading of inclination (or dip amount) which may be variable between horizontal (00) to vertical (900) is measured with the help of a mark in the middle of the frame opening. The bridge fitted outside the dial helps in taking reading of the inclination of bed. The bridge is movable which can rotate upto 900 angles from a plane parallel to the dial up to a plane perpendicular to the dial. The bridge line is always parallel to the North-South line of the dial. The locking device is present outside the dial ring which on pressing locks the magnetic needle as well as clinometer so that the reading may be noted even after removing the compass from outcrop.

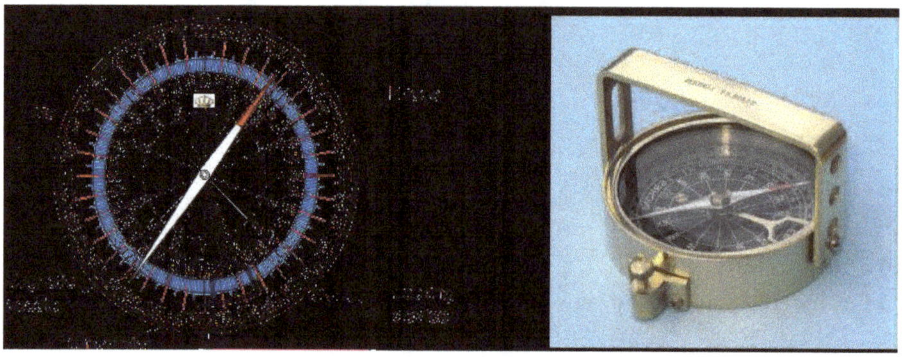

Sketch and photograph of clinometer compass.

Measuring Strike and Dip of an Inclined Bed

Let us discuss following procedure to measure the strike and dip of bed:

1. To measure the strike direction, place the compass on bedding plane in so that the bridge of the compass touches the bedding plane completely.

2. Then rotate the compass so as to ensure that the bridge becomes horizontal and one end of the bridge still touches the bedding plane. Now let the magnetic needle move freely. Let the needle come to rest and then read both ends of the azimuthal circle which represent strike.

3. In order to measure the amount of dip of the bed, draw a line on the bed perpendicular to strike and keep the bridge on the bedding plane along this line in such a way that the dial plane is vertical. The reading in clinometer gives the amount of dip.

4. In order to measure the direction of the dip, place the bridge along the line drawn on

the bedding plane so that the dial face the sky. Then rotate the compass to horizontal so that the bridge and N-S line of the dial both remain parallel to the line. The crown is often marked as N. Take care the crown in the dial is towards the dip direction of the bed.

5. On rotation of clinometer to the horizontal, the N marked end of the magnetic needle gives the dip direction. You can check your reading. The true dip and strike directions are always perpendicular to each other.

Rock Structures

Rock structures can be broadly divided into:

- Planar structures;
- Linear structures;
- Folds and faults

Planar Structures

Planar structures are those structures which are found as a plane or as a surface in the rocks. Planar structures go hand in hand with linear structures. Let us discuss major planar structures:

1. Bedding plane: They are the planes that bound a sedimentary bed. A bedding plane separates the two beds in a sedimentary rock.

2. Metamorphic foliation: It is a planar structure which is formed during deformation and metamorphism of a pre-existing rock. They are characteristic feature found in the metamorphic rocks such as slate, schist and gneiss.

3. Igneous foliation: They are observed with igneous rocks and are also known as primary foliation.

4. Fault plane: Fault plane is a fracture along which the rock blocks have been displaced.

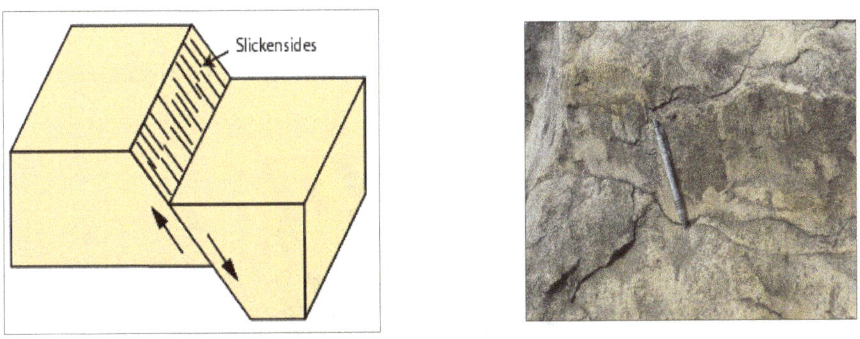

(a) (b)

(a) Displacement has taken place along the fault plane. Direction of movement is marked by arrows in opposite directions, and (b) Field photograph of slickensides.

5. Bedding plane: They are the planes that bound a sedimentary bed. A bedding plane separates the two beds in a sedimentary rock.

6. Metamorphic foliation: It is a planar structure which is formed during deformation and metamorphism of a pre-existing rock . They are characteristic feature found in the metamorphic rocks such as slate, schist and gneiss.

7. Igneous foliation: They are observed with igneous rocks and are also known as primary foliation.

8. Fault plane: Fault plane is a fracture along which the rock blocks have been displaced.

9. Joints: They are irregular or regular planar separated portion found in the rocks. Joints are the fracture surfaces along which movement is negligible and/or not observable. They are among the most common of all geological features. The systematic study of joints in an area can unravel the timing and sequence of its formation. The cooling of lava or magma can produce joints in the rocks. Fracture and joints are among the most important geological structures considered in hydrology, engineering, mining projects.

Rhombohedral joints in igneous rock.

Linear Structures

Let us discuss about the linear structures. As the name suggests these structures is line like features observed in the rocks.

1. Mineral lineation: If the longer dimensions of the minerals are aligned in a particular direction, it gives rise to mineral lineation.

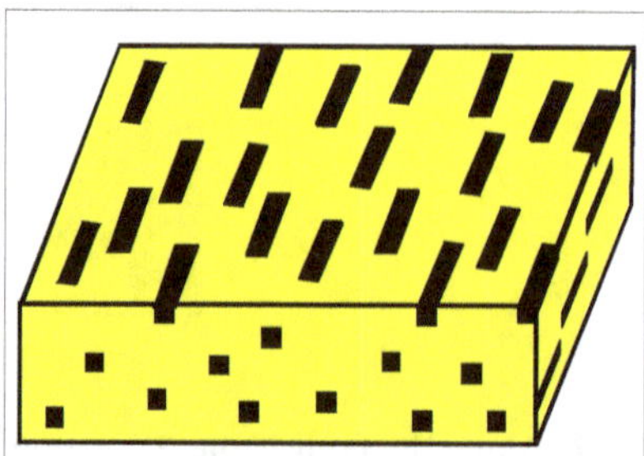

Mineral lineation

2. Crenulation lineation: They form when any planar surface such as foliation is affected by folding on small or microscopic scales so that hinge lines of these folds are aligned in a particular direction.

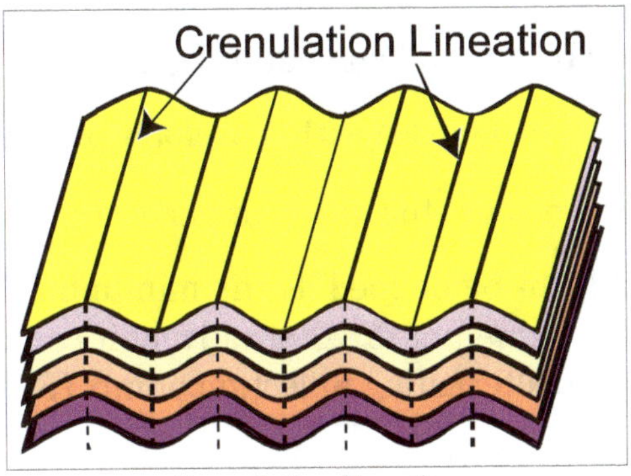

Crenulation lineation

3. Slickensides: They are lines like features developed on the fault plane because of friction generated between two faulted blocks and usually show polished surface.

4. Boudinage, pinch and swell and roddings: When the competent rock layers stretch and deforms into segments they form boudins. Individual boudins are commonly much longer in one dimension than other two. Roddings describes elongated mineral aggregates.

Folds and Faults

Now we will discuss in detail about folds and faults

Folds

Folds are combination of planar and linear structures. You can imagine folds in rocks are like folds in clothing. Similarly when the layers of rocks suffers gradual compression by the tectonic forces in the crust, they are pushed into folds. The layers of rocks can be crumpled or buckled into folds. Folding can be defined as the bending of rock strata due to compressional forces acting tangentially or horizontally towards a common point or plane from the opposite sides. We can define fold as the wave-like undulation wherein the bending or arching of the rock layers takes place due to forces of the Earth. Folding is a common form of deformation displayed in the layered rocks. Folds are best displayed by stratified formations, *viz.*, sedimentary or volcanic rocks or metamorphic rocks. Folds vary in size from centimetre on outcrop scale to hundreds of kilometers on the regional scale.

Importance of folding: The study of folds in geological studies is important because of following reasons:

- Folding exposes the deep seated rocks on the surface of the Earth.

- It increases the mineral deposits because of repetition of layers due to folding in a limited area.
- It facilitates development of site for deposition of mineral bearing solution.
- Folds serve as good host for oil and natural gas.
- Folding causes beautiful landscapes to develop which may enhance geotourism.

Parts of the fold: Now let us discuss the different parts or elements of a fold.

- Wavelength of fold can be defined as the minimum distance between its two successive points of same phase. Alternatively it can also be explained as the distance between two alternating inflection points (Figure).

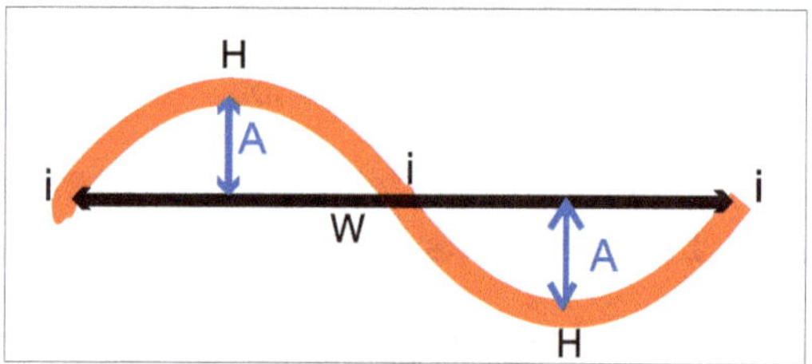

Parts of a fold (Abbreviations used: W-wavelength, A amplitude, H-hinge point, i-inflexion point

- Crest is the highest point in the profile section of the fold (Figure).
- Trough is the lowest point in the profile section of a fold (Figure).
- Crestal line is highest located line in the fold. It can be obtained by joining the crestal points in a folded layer (Figure).
- Trough line is obtained by joining the trough points of a folded layer (Figure). This line is located lowest in the fold.
- Culmination is the highest point located on the crestal line (Figure).
- Depression is the lowest point located on the trough line (Figure).

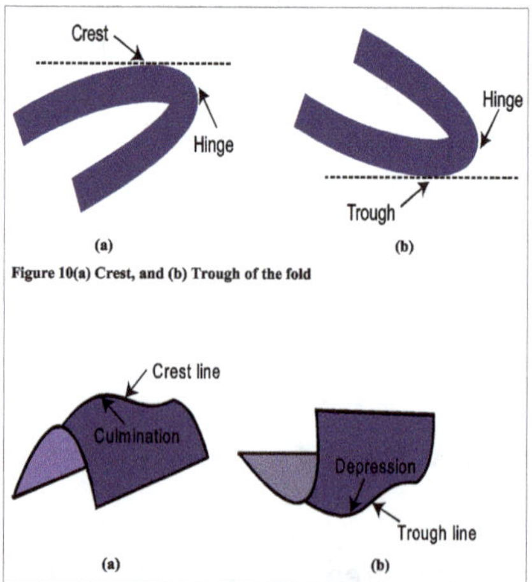

(a) Culmination, and (b) Depression.

- Fold axis is an imaginary line which by moving parallel to itself generates the fold. The hinge line of a fold is considered equivalent to 'fold axis' if it is straight and the fold is cylindrical in nature.
- Inflexion point is that point where the fold limb changes its attitude.
- Inflexion line is obtained by joining the inflexion points of a folded layer.
- Limb is a side between inflexion point and hinge.

(a)

Parts of fold (a) Sketch, and (b) Field photograph of a fold.

- Axial surface/ plane is formed by joining of fold hinge lines of successive beds. When the axial surface forms a plane it is called axial plane is formed by the axial surface.
- Amplitude of a fold is the length of perpendicular drawn from hinge point of the fold on the line joining the two inflection points of the fold.
- Hinge point It is the point of maximum curvature on the profile section of a fold. The profile of the fold is a cross section or transverse section across the hinge line of the fold.
- Hinge Zone: Sometimes the maximum curvature of the fold is not at a point but a set/group in a zone called hinge zone.
- Hinge line: It is the locus of hinge points of a particular bedding plane. Hinge or hinge line of a fold is the line of maximum curvature in the folded bed.

Faults

Now let us discuss about faults.

Fault can be defined as a fracture along which there is an observable amount of dislocation or displacement of the two rock blocks. Like folds, faults also occur in all sizes. The development of fault in rock takes place due to tectonic stresses such as tensional, tangential or compressional or in combination.

Importance of Faults: Faults play a significant role in geological studies and can pose challenge to geologists while mapping.

- Faulting exposes the rocks from the deeper level to the Earth's surface which provides knowledge of the subsurface geology.
- Faults provide the excellent channel for the movement of mineralized solution or petroleum. They trap petroleum from migration and loss.
- Faults may create beautiful landscapes which enhance geotourism.
- They may be the cause of origin of earthquakes.
- Fault locations are very important prospects in mining and exploration.

Field photograph of fault

Parts of Fault: Let us analyse different parts of fault.

- Fault surface/ fault plane: It is a surface or plane along which the dislocation takes place (Figure).
- Fault zone: It is a zone of numerous small scale fractures constituting fault (Figure).
- Fault line/ fault trace: It is the line formed by intersection of a fault with surface of the Earth or any given surface (Figure).
- Dip of the fault plane: It is the angle of inclination of the fault plane with respect to horizontal. The true dip is measured on a vertical section perpendicular to strike (Figure).
- Strike of the fault plane: It is the direction in which the fault plane cuts the horizontal ground and is strike of that plane (Figure).
- Hade of the fault plane: It is the angle made by fault plane with respect to a vertical line (Figure).

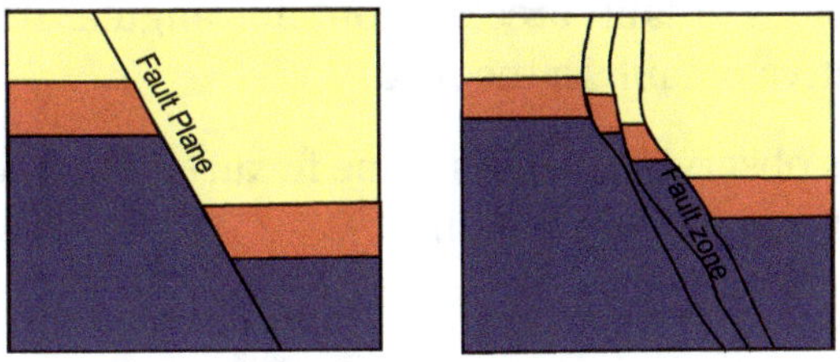

(a)
(b)
Figure 14(a) Fault plane, and (b) Fault zone

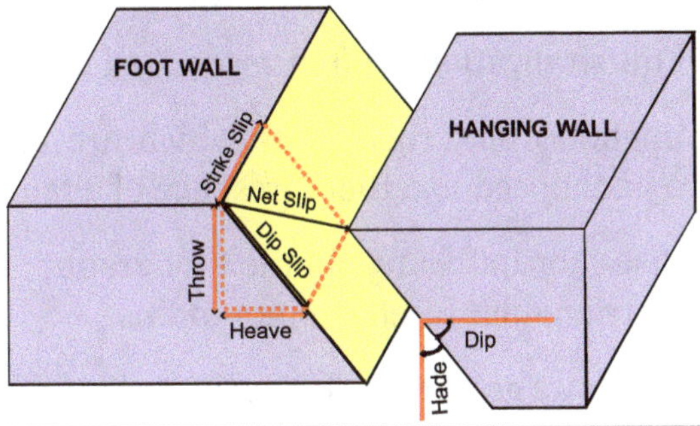

Parts of the fault3

- Hanging wall: The rock mass resting above the inclined fault plane is known as hanging wall (Figure).
- Foot wall: The rock mass resting below the inclined fault plane is known as foot wall (Figure).
- Slip: It refers to any displacement parallel to fault plane in the given direction (Figure).
- Net- Slip: It is the total amount of displacement measured parallel to fault plane (Figure).
- Dip-slip: It is the movement parallel to dip direction of the fault plane (Figure).
- Strike-slip: The movement parallel to strike of the fault plane (Figure).
- Oblique-slip: It is the movement called oblique-slip, if both dip-slip and strike-slip movements are present. (Figure).
- Heave: It is the horizontal component of the dip slip (Figure).
- Throw: It is the vertical component of the dip slip (Figure).

Recognition of Faults in the Field: Let us discuss features that are characteristically associated with fault plane. The presence of these features in the field is indicative of faulting.

- Visible displacements of veins, dikes, strata, *etc.* in the field are best evidence of faulting generally seen on the outcrop scale. For larger scale faulting geologists look for indirect evidences in the field.

- Presence of slickensides or fault breccia (contains angular fragments) along the fault plane is indicative of fault in the field.
- Dragging of strata is observed; there is some flexure or bending of beds present along a plane which may be a fault plane.
- Presence of crushing, shearing and pulverization of rock indicates faulting.
- Presence of mylonites (fine crushed rock materials) is indicative of fault zone.
- Silicification and mineralization is often associated with fault plane.
- Sedimentary or metamorphic rock types which are usually found at distant places, if they are juxtaposed together, presence of a fault may be indicated.
- Faulting often causes abrupt change in the topography, which can be observed in the form of cliff, triangular facets on mountains.
- If a river or stream shows an abrupt change in its flow direction or it makes waterfall; it is indicative of presence a fault.
- If the springs are found along a line, there may be presence of fault.

References

- Blatt, H.; Tracy, R.J. & Owens, B.E.; 2006: Petrology, Igneous, Sedimentary, and Metamorphic, W.H. Freeman & company, New York (3rd ed.), ISBN 978-0-7167-3743-8
- "Digital Cartographic Standard for Geologic Map Symbolization". FGDC Geological Data Subcommittee. USGS. August 2006. Retrieved 07, August 2020
- V. Guerriero; et al. (2011). "Improved statistical multi-scale analysis of fractures in carbonate reservoir analogues". Tectonophysics. Elsevier. 504: 14–24. Bibcode:2011Tectp.504...14G. doi:10.1016/j.tecto.2011.01.003
- B.A. van der Pluijm and S. Marshak (2004). Earth Structure - An Introduction to Structural Geology and Tectonics (2nd ed.). New York: W. W. Norton. p. 656. ISBN 0-393-92467-X
- V. Guerriero; et al. (2009). "Quantifying uncertainties in multi-scale studies of fractured reservoir analogues: Implemented statistical analysis of scan line data from carbonate rocks". Journal of Structural Geology. Elsevier. 32 (9): 1271–1278. Bibcode:2010JSG....32.1271G. doi:10.1016/j.jsg.2009.04.016

Chapter 5

Sedimentology: An Integrated Study

The scientific discipline that deals with the study of composition and texture of sedimentary rocks and the processes like transportation, deposition and lithification that are involved in the formation of sedimentary rocks is known as sedimentology. The aim of this chapter is to explore the topics related to sedimentology which are crucial for a complete understanding of the subject.

Sedimentology studies sand, mud (or silt) and clay, and the various ways they are deposited. Using these studies, sedimentologists apply their understanding of modern processes to ancient rock, to try to understand how it formed.

Most of the rocks on earth are sedimentary rocks, and it is in these kinds of rocks you find fossils and many of the other historical markers. Sedimentary rocks are also where petroleum deposits are found. Sedimentology is tied to stratigraphy, which studies the relationships between rock layers and how they can shift and move. This also affects where petroleum deposits can be found, as well as how the extraction of petroleum affects the sediment around the deposit.

Why Sedimentology Matters

Sediments and sedimentary rocks cover most of Earth, and weathering is occurring on the rest of it. The reshaping of the surface of the Earth has had a huge influence on the planet, affecting everything from the evolution of life to the tectonics of mountain ranges. Sediments and sedimentary rocks record the events and processes that shaped the surface of Earth – and other rocky planets. They provide the temporal framework that connects processes within the Earth to those at the surface. They are important for:

- Earth history: Sedimentary rocks contain features that allow us to interpret ancient depositional environments, including the evolution of organisms and the environments they lived in, how climate has changed throughout Earth history, where and when faults were active, etc.

- Economic resources: Petroleum reservoirs have organic-rich, sedimentary source rocks that produced the petroleum when heated, most oil and gas migrates through sedimentary rocks, and most of the reservoirs are hosted in sedimentary rocks. Water aquifers are dominantly found in sedimentary rocks (although some are in fractured metamorphic and igneous rocks). The composition of the rocks strongly influences water quality due to water-rock interactions. (Why does Davis water taste bad?) Sedimentary rocks also host economic minerals such as gold and diamonds, which are eroded from other rocks and concentrated to specific areas during sediment transport.

- Environmental geology: Sediments cover 2/3 of the continents and essentially all of the ocean floor, which totals 89% of the surface of Earth. They host the biosphere, and they are most of the rocks we interact with directly and indirectly. Our actions as humans have an extremely strong effect on sedimentation and erosion. Understanding our impact on the environment - and the environment's impact on us - must include deep appreciation for sediments and sediment transport.

Sediment Grain Size, Measurement Scales and Methods of Grain Size Measurement

Grain size represents a fundamental character of clastic sedimentary rock and one of the most important property for description of these rocks. Sizes of clastic grains reflect their weathering and erosional history, besides providing idea about their transportation motif i.e as bed load, saltation or suspension. By the term grain size normally we mean grain diameter. This is true when the grain is perfectly spherical. However, most grains in nature are non-spherical and that generates most relevant question 'what is the representative diameter for a non-spherical grain?' To resolve this question the concept of 'nominal diameter' is proposed. The 'nominal diameter' of a grain is the diameter of a sphere that has equal volume as that of the grain. Again, in practice we are seldom in a position to measure volume of each grain, and therefore use indirect method for measurement of any grain. If a grain resembles tri-axial ellipsoid, we measure its long axis (a), intermediate axis (b) and short axis (c) for further use of data.

As in nature grain size of sediments vary widely i.e from clay size less than 0.002mm in diameter to boulder size more than 25cm in diameter, necessity was felt to categorize grains in terms of size scale.

Grain Size Scales

- Udden-Wentworth Scale: Proposed by Udden in 1914, the scale divides consecutive classes of grain sizes in millimeter scale by multiples of 2 on higher side and decrease by multiple of 1/2 on the lower side.

Boulder I	Cobble I	Pebble I	Granule I	Sand I	Silt I	Clay
256	64	4	2	1/16	1/256	(mm)

As commonly available grain sizes in nature falls in fine-size category i.e sand to clay and their class intervals comes in fraction following this scale, computations became very cumbersome in early days in absence of any high speed computational facility. This gave birth to another grain size scale which is referred as Phi scale.

- Phi (Krumbein) Scale: Krumbein in 1934 devised this scale that is referred as phi (Φ) scale.

  Phi (Φ) = $-\log_2 d$; where d is diameter of a grain in mm

The use of 2 as the base of the logarithm allowed direct conversion of Udden's grade scale to Krumbein phi scale. Also, inversion of sign transformed the fractions to positive, round integers, thereby making computations much easier for the finer grains (sand, silt, clay).

Wentworth nomenclature	Boulder		Cobble		Pebble		Granule		Sand		Silt		Clay
Udden's grade scale		256 (2^8)		64 (2^6)		4 (2^2)		2 (2^1)		1/16 (2^{-4})		1/256 (2^{-8})	mm
Krumbein's phi-scale		-8		-6		-2		-1		+4		+8	ϕ

Conversion of Udden-Wentworth scale to Krumbein phi scale.

Measurement of Grain Size

Gravel and Sand Size

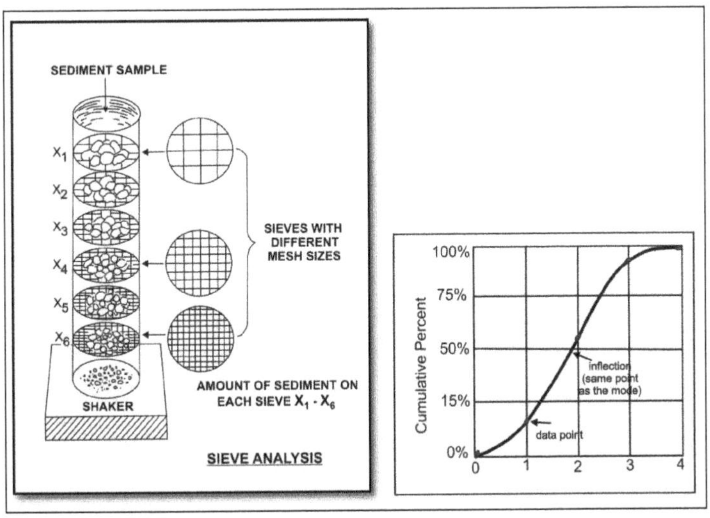

Bank of seives with decrease in opening size downward. Depending on number of seives used and their respective opening sizes, the original sediment sample get divided into a number of size fractions on seiving.

Measurement of sand and gravel grains are done by means of mechanical seiving. Seives are screens of standard size and specified by the American Society of Testing Materials (A.S.T.M). Openings of screens increase uniformly in multiples of $\sqrt[4]{2}$ following size grades of Wentworth size scale, starting from an opening of 0.0029 inch (200 mesh). For size measurement, a bank of seive is taken with downward decreasing opening (mesh) size. Sediment with measured weight is kept on the topmost seive and then the seive bank is mechanically or electrically shaken. Depending on the size of grains, the sediment get distributed among the seives used; grains larger than the seive size will remain on the seive and grains with smaller size pass on to the seives further down. By weighing fraction of sample lying on each seive, a grain size distribution curve is constructed. The lower limit of seive analysis is 0.04-0.03 mm, grains finer than this size show large amount of cohesion and do not allow measurement of individual grain by this method.

Silt and Clay Size Fraction

Sediments of this size fraction cannot be measured by seive analysis because of their high cohesive character. In lieu, measurement of size for these sediments are done through number of other methods, most of which deals with settling velocity of sediment grain in liquids following Stoke's law.

According to the Stoke's law when the settling velocity of grains falling through a uniform solution (solution at which density is same at all depth levels) is constant, the resistance to the movement (friction), which acts upward balance the force of gravity on the grain, which acts downward.

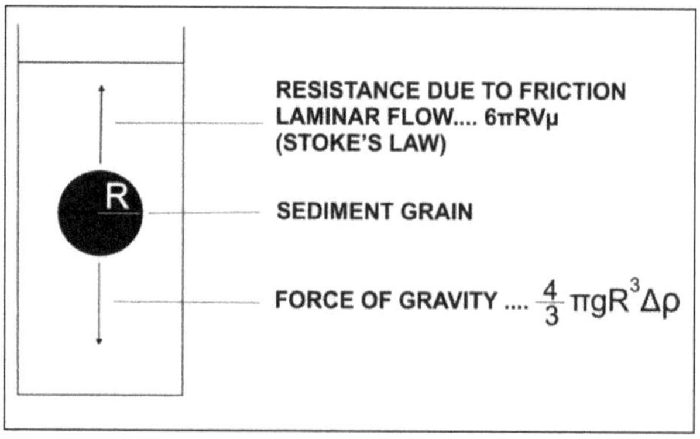

Forces act on a falling sediment grain. For a grain settling with uniform velocity the downward force of gravity balance upward frictional force.

$$\underset{\text{(Friction)}}{6\pi R v \mu} = \underset{\text{(Gravity)}}{4/3\, \pi g R^3 \Delta\rho}$$

$$v = 2/9\, \mu g\, R^2\, \Delta\rho$$

Where μ is viscosity and Δρ is density difference between the grain and liquid. R is the radius (in cm) of sediment grain. From the relation it can be noticed that the settling velocity proportionally varies with the radius of the grain ($V \propto R^2$) as π, g and $\Delta\rho$ are constant between the grain and liquid. For settling through water, the relation becomes $\log v = 2 \log R + C$.

This relationship applies well for grains smaller than 0.1mm in diameter. With increase in grain size (i.e 0.1 to 0.5mm) settling velocity increases and turbulence develops around the grain. Settling velocity of such relatively large size grains varies proportionally with square root radius of the grain.

However, measurement of settling velocity of each grain is also not a practicable idea and hence, attempts were made to overcome the difficulty in some indirect way by use of hydrometer or sedimentation balance. Other methods which are in use in recent times include refraction or dispersion of light through suspensions of sediment (scattering) or use of x-ray instead of light.

Hydrometer

A suspension of sample is created with a mixer in a cylinder so that at time T_0 there is an even distribution of grains in the solution and density of solution is same throughout the cylinder. A hydrometer will float in the solution and register density of the solution on a scale on the upper part of the tube. If the suspension is allowed to stand for a time T1, grains will settle in the bottom of the cylinder. At the level at which the bulb of the hydrometer is floating the largest of the grain sizes already must have settled resulting decrease in density of the solution. This reduction in density of solution gets registered by further submergence of hydrometer and recorded in the scale at the upper part of tube. By taking successive readings of the density we plot a density variation curve with respect to time. Since density reduction is a function of settling velocity, the density variation curve can be calibrated as a grain-size distribution curve. In case the sediment contains a large clay fraction, use of centrifuge may be a more desired option. The acceleration term g in the Stoke's law increased in the centrifuge, which can be calculated from the velocity of rotation and length of rotating arm.

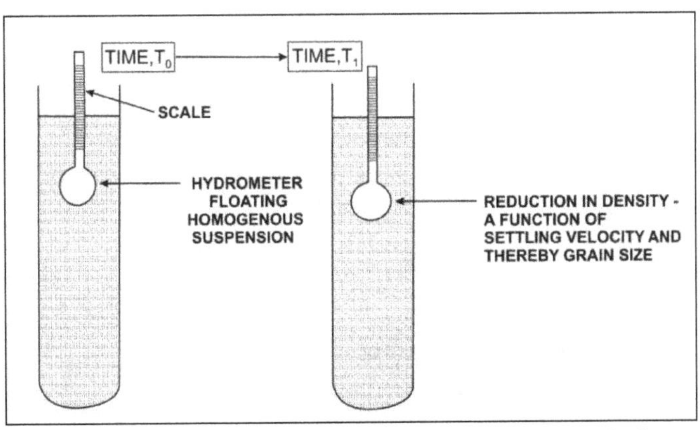

Operational procedure of a hydrometer. Note change in degree of submergence of hydrometer in a solution with time depending on change in density of solution.

Sedimentation Balance

In this method sediments suspended in a solution in a cylinder fall through the water column and settle on a balance pan kept at any depth level within the cylinder. The balance pan is kept connected with a weigh meter. With passing time more and more sediment will get settled on the pan and the increasing weight of the pan get recorded in the weigh meter. The increase in weight is plotted against time, which indicated suspension settlement from solution as a function of time. This gives a direct cumulative curve that can be calibrated to give a grain-size distribution curve.

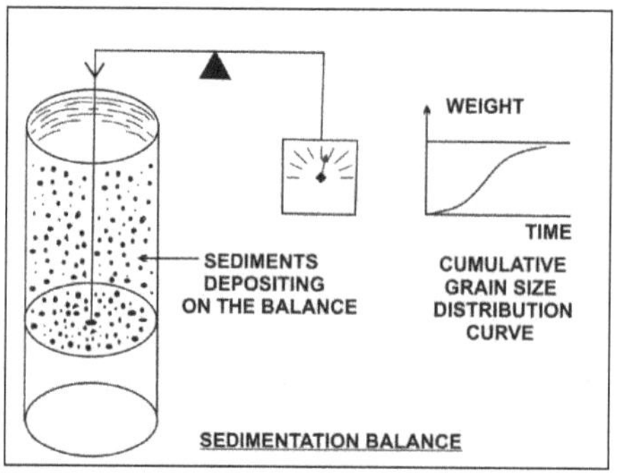

Use of sedimentation balance for measurement of increase of weight with respect to time and therefore cumulative grain size distribution.

However, settling velocity of particles depend on their shapes. A spherical grain settles faster than a non-spherical grain of same mass. Hence, measurement of size for non-spherical grains through sedimentation balance may not yield same result as that from the seiving method.

Grain Size Measurement in Rocks

Grain size measurement in sedimentary rocks is done in thin section by use of microscope. However, it is difficult to measure fine fractions (fine silt and clay) in this method. One major caution in this method is 'sectional effect' i.e under thin section diameter of a grain will depend on relative orientation of grain and the cut section on which thin section is made and hence, in most cases we do not get the greatest diameter. With spherical grains, statistically expressed relation between real diameter d_r and observed diameter d_o is $d_r = 4/\pi d_o$. However, in practice we do not give this correction to measured grain sizes from a rock as sectional effect on all grains of a rock makes no change on their relative sizes.

Sedimentary Structures

Sedimentary structures are visible textures or arrangements of sediments within a rock. Geologists use these structures to interpret the processes that made the rock and the environment in which it formed. They use uniformitarianism to usually compare sedimentary structures formed in modern environments to lithified counterparts in ancient rocks.

Bedding Planes

Horizontal strata in southern Utah.

The most basic sedimentary structure is bedding planes, the planes that separate the layers or strata in sedimentary and some volcanic rocks. Visible in exposed outcroppings, each bedding plane indicates a change in sediment deposition conditions. This change may be subtle. For example, if a section of underlying sediment firms up, this may be enough to create a form or a layer that is dissimilar from the overlying sediment. Each layer is called a bed, or stratum, the most basic unit of stratigraphy, the study of sedimentary layering.

Students from the University of Wooster examine beds of Ordovician limestone in central Tennessee.

As would be expected, bed thickness can indicate sediment deposition quantity and timing. Technically, a bed is a bedding plane thicker than 1 cm (0.4 in) and the smallest mappable unit. A layer thinner than 1 cm (0.4 in) is called a lamina. Varves are bedding planes created when laminae and beds are deposited in repetitive cycles, typically daily or seasonally. Varves are valuable geologic records of climatic histories, especially those found in lakes and glacial deposits.

Graded Bedding

Image of the classic Bouma sequence. A=coarse- to fine-grained sandstone, possibly with an erosive base. B=laminated medium- to fine-grained sandstone. C=rippled fine-grained sandstone. D=laminated siltstone grading to mudstone.

Graded bedding refers to a sequence of increasingly coarse- or fine-grained sediment layers. Graded bedding often develops when sediment deposition occurs in an environment of decreasing energy. A Bouma sequence is graded bedding observed in a clastic rock called turbidite. Bouma sequence beds are formed by offshore sediment gravity flows, which are underwater flows of sediment. These subsea density flows begin when sediment is stirred up by an energetic process and becomes a dense slurry of mixed grains. The sediment flow courses downward through submarine channels and canyons due to gravity acting on the density difference between the denser slurry and less dense surrounding seawater. As the flow reaches deeper ocean basins it slows down, loses energy, and deposits sediment in a Bouma sequence of coarse grains first, followed by increasingly finer grains.

Flow Regime and Bedforms

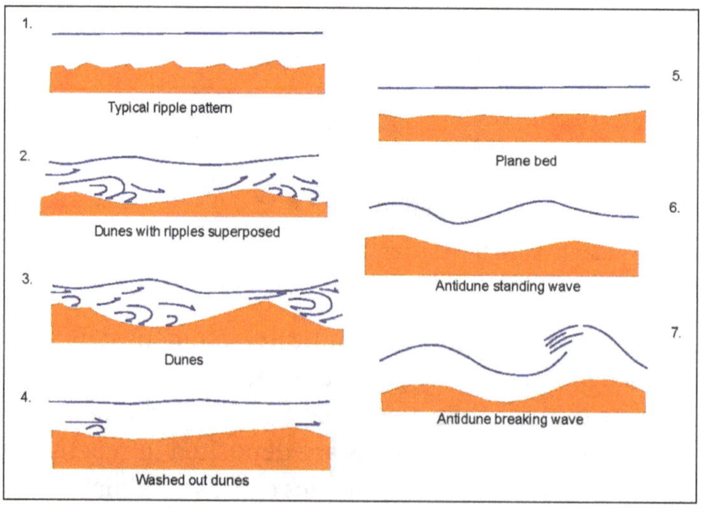

Bedforms from under increasing flow velocities.

In fluid systems, such as moving water or wind, sand is the most easily transported and deposited sediment grain. Smaller particles like silt and clay are less movable by fluid systems because the tiny grains are chemically attracted to each other and stick to the underlying sediment. Under higher flow rates, the fine silt and clay sediment tend to stay in place and the larger sand grains get picked up and moved.

Bedforms are sedimentary structures created by fluid systems working on sandy sediment. Grain size, flow velocity, and flow regime or pattern interact to produce bedforms having unique, identifiable physical characteristics. Flow regimes are divided into upper and lower regimes, which are further divided into uppermost, upper, lower, and lowermost parts. The table below shows bedforms and their associated flow regimes. For example, the dunes bedform is created in the upper part of the lower flow regime.

Flow Regime (part)	Bedform	Description
Lower (lowest)	Plane bed	Lower plane bed, flat laminations.
Lower (lower)	Ripples	Small (with respect to flow) inclined layers dipping downflow.
Lower (upper)	Dunes	Larger inclined cross beds, ±ripples, dipping downflow.
Upper (lower)	Plane bed	Flat layers, can include lined-up grains (parting lineations).
Upper (upper)	Antidunes	Hard to preserve reverse dunes dipping shallowly upflow.
Upper (uppermost)	Chutes/pools *(rare)*	Erosional, not really a bedform; rarely found preserved.

Plane Beds

Subtle lines across this sandstone (trending from the lower left to upper right) are parting lineations.

Plane beds created in the lower flow regime are like bedding planes, on a smaller scale. The flat, parallel layers form as sandy sediment piles and move on top of layers below. Even non-flowing fluid systems, such as lakes, can produce sediment plane beds. Plane beds in the upper flow regime are created by fast-flowing fluids. They may look identical to lower-flow-regime beds; however, they typically show parting lineations, slight alignments of grains in rows and swaths, caused by high sediment transport rates that only occur in upper flow regimes.

Ripples

Modern current ripple in sand from the Netherlands. The flow creates a steep side down current. In this image, the flow is from right to left.

Ripples are known by several names: ripple marks, ripple cross-beds, or ripple cross laminations. The ridges or undulations in the bed are created as sediment grains pile up on top of the plane bed. With the exception of dunes, the scale of these beds is typically measured in centimeters. Occasionally, large flows like glacial lake outbursts can produce ripples as tall as 20 m (66 ft).

A bidirectional flow creates this symmetrical wave ripple. From rocks in Nomgon, Mongolia. Note the crests of the ripples have been eroded away by subsequent flows in places.

Climbing ripple deposit.

First scientifically described by Hertha Ayrton, ripple shapes are determined by flow type and can be straight-crested, sinuous, or complex. Asymmetrical ripples form in a unidirectional flow. Symmetrical ripples are the result of an oscillating back-and-forth flow typical of intertidal swash zones. Climbing ripples are created from high sedimentation rates and appear as overlapping layers of ripple shapes.

Dunes

Lithified cross-bedded dunes from the high country of Zion National Park, Utah. The complexity of bedding planes results from the three-dimensional network of ancient dune flows.

Dunes are very large and prominent versions of ripples and typical examples of large cross-bedding. Cross bedding happens when ripples or dunes pile atop one another, interrupting, and/or cutting into the underlying layers. Desert sand dunes are probably the first image conjured up by this category of bedform.

British geologist Agnold considered only Barchan and linear Seif dunes as the only true dune forms. Other workers have recognized transverse and star dunes as well as parabolic and linear dunes anchored by plants that are common in coastal areas as other types of dunes.

The modern sand dune in Morocco.

Dunes are the most common sedimentary structure found within channelized flows of air or water. The biggest difference between river dunes and air-formed (desert) dunes is the depth of the fluid system. Since the atmosphere's depth is immense when compared to a river channel, desert dunes are much taller than those found in rivers. Some famous air-formed dune landscapes include the Sahara Desert, Death Valley, and the Gobi Desert.

As airflow moves sediment along, the grains accumulate on the dune's windward surface (facing the wind). The angle of the windward side is typically shallower than the leeward (downwind) side, which has grains falling down over it. This difference in slopes can be seen in a bed cross-section and indicates the direction of the flow in the past. There are typically two styles of dune beds: the more common trough cross-beds with curved windward surfaces, and rarer planar cross-beds with flat windward surfaces.

In tidal locations with strong in-and-out flows, dunes can develop in opposite directions. This produces a feature called herringbone cross-bedding.

Herringbone cross-bedding from the Mazomanie Formation, upper Cambrian of Minnesota.

Hummocky-cross stratification, seen as wavy lines throughout the middle of this rock face. The best example is just above the pencil in the center.

Another dune formation variant occurs when very strong, hurricane-strength, winds agitate parts of the usually undisturbed seafloor. These beds are called hummocky cross-stratification and have a 3D architecture of hills and valleys, with inclined and declined layering that matches the dune shapes.

Antidunes

Antidunes are so named because they share similar characteristics with dunes, but are formed by a different, opposing process. While dunes form in lower flow regimes, antidunes come from fast-flowing upper flow regimes. In certain conditions of high flow rates, the sediment accumulates upstream of a subtle dip instead of traveling downstream. Antidunes form in phase with the flow; in rivers, they are marked by rapids in the current. Antidunes are rarely preserved in the rock record because the high flow rates needed to produce the beds also accelerate erosion.

Antidunes forming.

Bioturbation

Bioturbated dolomitic siltstone.

Bioturbation is the result of organisms burrowing through soft sediment, which disrupts the bedding layers. These tunnels are backfilled and eventually preserved when the sediment becomes rock. Bioturbation happens most commonly in shallow, marine environments, and can be used to indicate water depth.

Mudcracks

Lithified mud cracks.

Mudcracks occur in clay-rich sediment that is submerged underwater and later dries out. Water fills voids in the clay's crystalline structure, causing the sediment grains to swell. When this waterlogged sediment begins to dry out, the clay grains shrink. The sediment layer forms deep polygonal cracks with tapered openings toward the surface, which can be seen in profile. The cracks fill with new sediment and become visible veins running through the lithified rock. These dried-out clay beds are a major source of mud chips, small fragments of mud or shale, which commonly become inclusions in sandstone and conglomerate. What makes this sedimentary structure so important to geologists is that they only form in certain depositional environments—such as tidal flats that form underwater and are later exposed to air. Syneresis cracks are similar in appearance to mud cracks but much rarer; they are formed when subaqueous (underwater) clay sediment shrinks.

Sole Marks

This flute cast shows a flow direction toward the upper right of the image, as seen by the bulge sticking down out of the layer above. The flute cast would have been scoured into a rock layer below that has been removed by erosion, leaving the sandy layer above to fill in the flute cast.

Sole marks are small features typically found in river deposits. They form at the base of a bed, the sole, and on top of the underlying bed. They can indicate several things about the deposition conditions, such as flow direction or stratigraphic up-direction. Flute casts or scour marks are grooves carved out by the forces of fluid flow and sediment loads. The upstream part of the flow creates steep grooves and downstream the grooves are shallower. The grooves subsequently become filled by overlying sediment, creating a cast of the original hollow.

Groove casts at the base of a turbidite deposit.

Formed similarly to flute casts but with a more regular and aligned shape, groove casts are produced by larger clasts or debris carried along in the water that scrape across the sediment layer. Tool marks come from objects like sticks carried in the fluid downstream or embossed into the sediment layer, leaving a depression that later fills with new sediment.

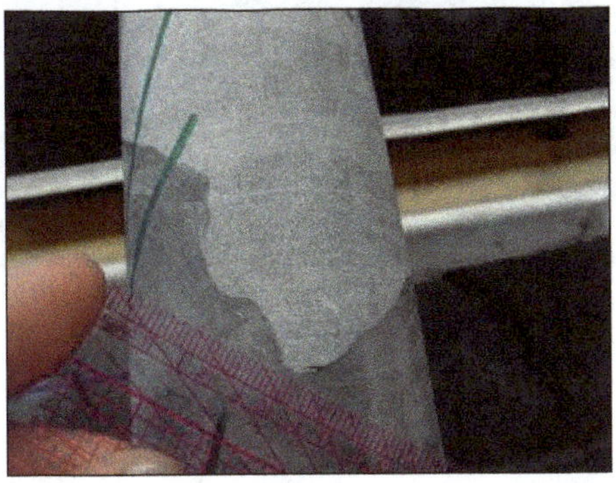

A drill core showing a loaded cast showing light-colored sand sticking down into dark mud.

Load casts, an example of soft-sediment deformation, are small indentations made by an overlying layer of coarse sediment grains or clasts intruding into a softer, finer-grained sediment layer.

Raindrop Impressions

Mississippian raindrop impressions over wave ripples.

Like their name implies, raindrop impressions are small pits or bumps found in soft sediment. While they are generally believed to be created by rainfall, they may be caused by other agents such as escaping gas bubbles.

Imbrication

Imbrication is a stack of large and usually flat clasts—cobbles, gravels, mud chips, etc—that are aligned in the direction of fluid flow. The clasts may be stacked in rows, with their edges dipping down and flat surfaces aligned to face the flow. Or their flat surfaces may be parallel to the layer and long axes aligned with the flow. Imbrications are useful for analyzing paleocurrents, or currents found in the geologic past, especially in alluvial deposits.

Cobbles in this conglomerate are positioned in a way that they are stacked on each other, which occurred as flow went from left to right.

Geopetal Structures

This bivalve (clam) fossil was partially filled with tan sediment, partially empty. Later fluids filled in the fossil with white calcite minerals. The line between the sediment and the later calcite is paleo-horizontal.

Eubrontes trace fossil from Utah, showing the geopetal direction is into the image.

Geopetal structures, also called up-direction indicators, are used to identify which way was up when the sedimentary rock layers were originally formed. This is especially important in places where the rock layers have been deformed, tilted, or overturned. Well-preserved mud cracks, sole marks, and raindrop impressions can be used to determine up direction. Other useful geopetal structures include:

- Vugs: Small voids in the rock that usually become filled during diagenesis. If the void is partially filled or filled in stages, it serves as a permanent record of a level bubble, frozen in time.
- Cross bedding: In places where ripples or dunes pile on top of one another, where one cross bed interrupts and/or cuts another below, this shows a cross-cutting relationship that indicates up direction.
- Ripples, dunes: Sometimes the ripples are preserved well enough to differentiate between the crests (top) and troughs (bottom).
- Fossils: Body fossils in life position, meaning the body parts are not scattered or broken, and trace fossils like footprints can provide an up direction. Intact fossilized coral reefs are excellent up indicators because of their large size and easily distinguishable top and bottom. Index fossils, such as ammonites, can be used to age date strata and determine up direction based on relative rock ages.
- Vesicles: Lava flows eliminate gas upwards. An increase of vesicles toward the top of the flow indicates up.

Sedimentary Environment

A sedimentary environment is an area of the earth's surface where sediments get deposited and can be separated physically, chemically and biologically from adjacent terrains. A sedimentary environment may be a site of erosion, nondeposition or deposition. Accordingly, environments can be dominantly Erosional or Depositional in nature. Subaerial environments are likely to be erosional while subaqueous environments are mostly depositional. Erosional environments not likely to be preserved in stratigraphic record and hence their existence can only be inferred from the detrital materials. Non-depositional or Equilibrial environment occur on land and sea, where transportation appears to be the dominant mechanism. Equilibrial environments do not deposit any detritus. They are preserved in the stratigraphic record as unconformities. The third type of sedimentary environment is depositional, which usually get preserved in stratigraphic record.

The best way to identify an ancient sedimentary environment is to relate them with their modern counterparts. Sedimentary processes operate on modern environments and gives rise to sedimentary products. Ancient environments contain the products, in form of rocks. Considering more or less similar processes operate within similar environments throughout the geologic time, it is easy to characterize each sedimentary environment on basis of their characteristics products (sediments or rocks). Ancient sedimentary (depositional) environments, hence, can be extrapolated from the rock records by comparing the products with their modern counterparts. However, such reconstruction is difficult and often has several inconsistencies. The process becomes harder as association of different depositional environment can often vary.

Major types of criteria used in recognizing sedimentary environments include the physical, chemical, and biological characteristics of the preserved rock mass. These are features that may be determined from a single outcrop or from several basinwide spanning outcrops or even subsurface cores. Where a larger number of samples are available, criteria of a much larger magnitude namely, the lateral and vertical facies relationships and three dimensional geometric framework of the strata can be employed. The physical characteristics that are of principal value include bedding characteristics, the nature of formation contacts, sedimentary structures and directional properties. Chemical analysis concentrates on the gross composition of the rock. The major mineral constituents and authigenic minerals can be especially useful in environmental reconstruction. Evidence of solution or accumulation of nodules allows further insight. Organic materials play a major role in the overall analysis. The floral and faunal assemblages, together with relative abundances and ratios of the various forms, are very useful. The lateral and vertical facies relationships and the three dimensional framework can greatly strengthen the environmental interpretation.

Ancient sedimentary environments can be characterized by deposits of a genetically related succession of laterally and vertically adjacent sedimentary rocks which accumulated under a specific set of environmental conditions. Sedimentary rocks record the geological development of these environments through both space and time. Physical primary structures, although most important, cannot predict the environment of their deposition alone. There are some unique sedimentary structures, which are environment-specific; the majority of them can form within various environments. Primary structures predict the mechanisms of their formation. Hence, these structures can easily be related with the processes operating within particular environment. Sedimentary rocks, classified into depositional facies, chiefly based upon these primary structures, actually represent the product of various processes operating on a particular depositional environment. After recognition of these structures, their association and distribution along time and space can give ample indications to identify depositional environment. This systematic approach is known as facies analysis.

Facies Analysis

The term Facies is introduced by Gressly from the Latin for 'aspect or appearance'. Reading defined facies as "a body of rock with specified characteristics". But the commonest definition of facies runs as "A body of rock characterized by a particular combination of lithology, physical and biological structures that bestow an aspect different from the bodies of rock above, below and laterally adjacent". No two sedimentary facies are ever identical. However, the term "facies" can be used in many senses. It is used as a strict observational sense of a rock product (sandstone facies), in a genetic sense for the products of a process by which a rock is thought to have formed (turbidite facies), in an environmental sense for the environment in which a rock is thought to have formed (fluvial facies), and as a tectofacies (postorogenic facies). All these uses are acceptable, as long as one remembers which concept is being used. Most informative among these four varieties are the genetic facies as it provides glimpses of formative mechanisms. Next comes the descriptive one. The classical sedimentary facies, or genetic facies, is properly defined on the basis of sedimentary structures present within the rock unit, along with their lithology, texture, body geometries etc. Description of a facies must highlight the uniqueness of the facies in the association. Subfacies incorporate the minor variations within the facies. For environmental interpretation though, a single facies is inadequate. Association of a group of facies and their distribution in time and space is required to understand the depositional environment properly.

Facies Association and Facies Succession

For meaningful depositional environmental or paleogeographic extrapolation, facies are interpreted in their association. Facies can commonly be grouped into facies associations, which can be defined as "groups of facies, genetically related to one another, and with some environmental significance". Facies associations are product of a particular sedimentation dynamics. Sedimentation dynamics, as a whole, is different between different sedimentary environments, but individual components are often shared. Only thorough familiarity with facies interrelationship one can classify different Facies Associations. Associations are the basic architectural elements or distinctive paleogeomorphic components in an environmental product.

Facies, or rather, facies associations occur in distinctive successions, characteristically known as facies successions, in which one or more defining parameter changes progressively up-section; like coarsening upward and thickening upward successions. Successions imply that certain facies properties change progressively in a specific lateral or vertical direction. That property, in turn, throws some light about the lateral as well as temporal changes in depositional environment; like a coarsening upward succession indicates an increase in flow power, as flow energy primarily controls the grain-size. Sedimentary successions are often made up of repeating cycles, where each cycle represents a particular facies association. Any one cycle is unlikely to contain all the facies and also in the correct sequence. An idealized facies association can be distilled from analysis of many genetically related cycles.

Apart from environmental interpretations, facies association is always helpful for paleogeographic interpretation, that is, geographic distribution of facies in space. Logical assumptions about the trend of deviation from the ideal association enable one to predict the paleogeography of an area. Body fossil and trace fossils also provide important clues. Trace fossils are more important in this regard, as they never suffer any postmortem transport and generally get accentuated, instead of being damaged like body fossils, during diagenesis.

According to the 'Law of the Correlation of Facies' developed by Walther facies or facies association contacts are very much important for establishing the relationship between depositional systems in space and the resulting stratigraphic succession. While gradational (vertical) transitions from one facies to another indicate original adjacency and genetic relationship during formation, sharp/erosional (vertical) contacts provides no evidence of contemporaneous genetic relationship of depositional environments, rather indicate juxtaposition of unrelated facies.

The systematic classification scheme, to evaluate the nature of sedimentary rocks in terms of sedimentation processes and to interpret the depositional setting therefrom, is known as facies analysis. The process involves classification of a rock assemblage in a way that unravels changes in paleogeography and /or depositional framework through space and time. Facies Analysis is the fundamental prerequisite for:

- Basin analysis.
- Extrapolation of the external (Extrabasinal) as well as the internal (Intrabasinal, in case of carbonate) causes producing the facies changes, gradual or abrupt.
- Successful implementation of Sequence stratigraphy, particularly in outcrop.
- Extracting an idealized facies model to characterize each depositional environment by combining the features of recent sediments as well as ancient sedimentary rocks.

Facies Model

A facies model is a general summary of a sedimentary environment, ideally presented in terms of their idealized facies distribution in time and space that make it applicable to different geological study areas. Facies models represent a distillation of the key facts, which together characterize a specific depositional system. As sedimentary systems are dynamic and vary through times, a good facies model embodies a large amount of information from numerous examples of the same depositional system, by studying both ancient and modern examples to identify concurrent themes. Facies models are thus prepared for following three purposes:

- To sum up general characteristics of products of different sedimentary environments inferred from the study area.
- To distil out universally applicable models for different sedimentary environments.
- To utilize the universal model/s as predictive tools in the study in a new area.

Glacial Environment

Glacial environments are defined as those where ice is a major transport process. Liquid water and wind can also transport sediment in these environments. Wind transport is common when there is little vegetation. Liquid water transport occurs when the ice melts.

The high viscosity of ice makes all ice transport of sediment laminar. Thus, grain sizes are not sorted. All of the sediment is transported together, with the ice, and it is deposited when the ice melts. There are several features that are characteristic of glacial environments, including the process of erosion.

Erosion

Erosion in glacial environments is dominated by physical processes:

- Ice freezing in cracks in rocks, breaking them up.
- Flow of glaciers "plucking" rocks up from the base of the flow.
- Grinding of rocks against each other and against the floor of the glacial valley as the ice flows.

These processes produce some distinctive sedimentary features including:

- Facetted clasts, e.g. rocks with smoothed off faces from dragging against other rocks.
- Striations and grooves in rocks from dragging against other rocks.
- Flat valley floors called glacial pavements that are smoothed off due to glacial flow.
- Rock flour, which is clay size lithic grains formed from the bits of rock that are abraded off as facets, striations, grooves, and glacial pavements form.

There is often little chemical weathering in glacial environments because temperatures are cold.

Deposition

Ice flows are laminar because they have very high viscosity. This can be seen in the ice cliffs along the edges of glaciers in Taylor Valley, Antarctica. The ice is particularly cold and is so viscous that it does not flatten out on the time scale of at least dozens of years. Because the flow is laminar, when the ice melts or sublimates, it dumps all grain sizes into one deposit, forming a diamictite. If one knows that the diamictite was deposited by ice, it is then called till or tillite. If the glacier melts on land, it leaves piles of till in moraines. If it melts over water, the debris is deposited into the water, commonly forming a till sheet. If only a few large clasts are deposited in the water, they are called "drop stones". These commonly are deposited by melting ice bergs that carry large grains out over lakes or the ocean, where they are deposited in (nearly) standing water.

As glaciers melt over land, melt water commonly reworks glacial till into braided river deposits. In arid environments, much less reworking of the sediment takes place.

Glacial Facies Assemblage

Morraine Facies

- Composed of diamictite; no sedimentary structures.
- Diamictite in mounds.
- Clast composition mostly lithic fragments, including silt and clay-sized rock flour.
- Clasts mostly angular, some with facets and striations.

Till Sheet Facies

- Composed of diamictite; sedimentary structures suggestive of turbidites in rare sandstone interbeds.
- Diamictite in sheets with rare shale and sandstone interbeds.
- Clast composition mostly lithic fragments, including silt and clay-sized rock flour.
- Clasts mostly angular, some with facets and striations.

Distal Glacial-Lacustrine/Marine Facies

- Shale with isolated large clasts and sandstone interbeds with sedimentary structures suggestive of turbidites.
- Clast composition mostly lithic fragments.
- Clasts mostly angular.
- Frequency of large clasts decreases away from the glacier.

Eolian Environment

Aeolian or eolian environments are those in which sediment deposition is primarily governed by wind. They are typically associated with arid environments (areas that recieve <250 mm/yr precipitation), for two main reasons. First, water is a much more powerful transport agent than air due to its fluid properties, so a lack of water is prerequisite for air to be the dominant sedimentary force. Second, lack of water leads to scarce vegetation, which further facilitates wind-driven erosion because vegetation can stabilize sediment and prevent it from being easily moved around.

Thus the most common eolian environments are rocky deserts and ergs (the technical term for sandy deserts). Another, less common, eolian environment includes areas with glacial outwash, where melting ice has released loose sediment, and plants have not yet had time to stabilize it.

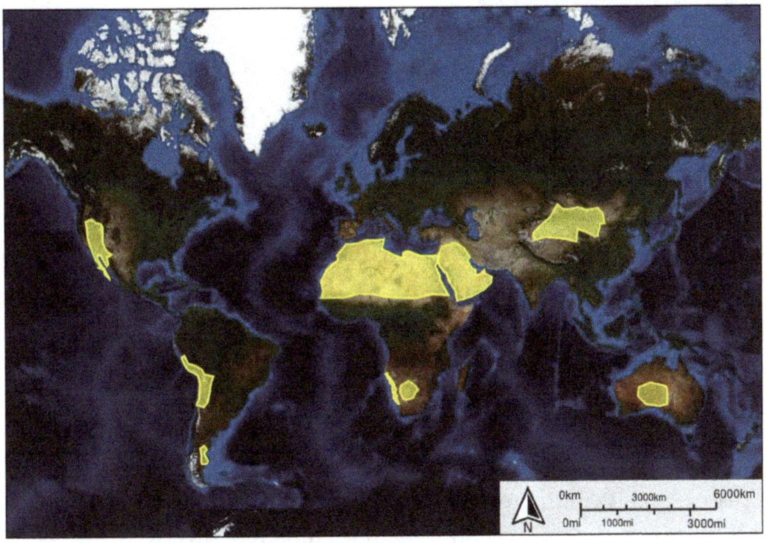

Distribution of the world's largest deserts.

Sediment Transport Processes

Ripples in an ice-age era loess deposit, where the grains have been compacted together.
Loess deposits can preserve important information about climate change over the last few million years.

Air has very different fluid properties compared to water, a fact which plays an important role in the eolian deposition of sediment. Air is orders of magnitude less dense than water, making grain transport much less efficient - in fact, it takes winds in excess of

60 mph just to move a grain 0.5 mm in diameter. Most sand grains are transported by rolling or saltation, though clay- and silt-sized particles can be carried in suspension by strong enough winds, sometimes in quantities great enough to cause dust storms. In areas where winds weaken, they may deposit their silt-sized suspended load in sequences known as loess. Air also has much lower viscosity than water, so air flows are generally turbulent.

Eolian Sedimentary Deposits

Composition and Grain Size

Air is much less dense than water, making it less capable of transporting large grains. Thus one characteristic of eolian deposits is that they are unlikely to contain sediment grains larger than coarse sand. Air's low viscosity also means that it has much less of a 'dampening' effect on impacts between grains, so the processes of abrasion that lead to a compositionally mature deposit are accelerated relative to water-driven processes. As a result, most aeolian deposits will have primarily well-rounded grains of quartz and perhaps chert, with little to no micas and weaker minerals. These grain impacts will also characterize their environment. In rocky deserts, rocks subjected to frequent, heavy wind conditions often have smooth, polished faces due to this sand-blasting effect.

Structures/Bedforms

Like aqueous processes, eolian processes generate familiar structures like ripples and dunes. They also create draas, which are like massive dunes hundreds of meters in wavelength and amplitude (draas are to dunes as dunes are to ripples, essentially).

Ripples on dunes in Death Valley National Park.

Eolian structures do differ slightly from their aqueous counterparts, however. Because air flow is very turbulent, there is no laminar flow sublayer like in water ripples. Instead, saltating grains push coarser grains up the back sides of the ripples to the crests where they eventually avalanche off the lee slope. The smaller grains at the crests, where wind speeds are highest, are transported away, and may accumulate in troughs between the larger grains. Thus large grains become more concentrated at the tops of ripples while smaller grains are more concentrated at the bases. Thin, reversely graded layers are therefore indicative of aeolian transport.

Eolian ripples and dunes migrate in the same way that aqueous ripples and dunes do, by continuous erosion on the windward side and deposition on the leeward side of these structures. Migrating dunes commonly create large-scale cross-stratification in the rock record, however migrating ripples rarely leave clear cross-lamination because the sand in these deposits is so well-sorted.

Facies

Arid environnments typically only have a few types of deposits in them. Aside from the aeolian sand dunes, these environments may contain poorly-sorted alluvial fan and river deposits, as strong but temporary flash floods are not uncommon in desert settings and can generate debris flows. These might manifest in the rock record as matrix -supported conglomerates, perhaps separated by thin beds of sandstone deposited between storm events. Arid environments may also support ephemeral lakes, which leave behind thin beds of mudstone and evaporite deposits when they dry up.

Lacustrine Environment

Lacustrine environments are insignificant in volume or areal extent among other depositional environments, but important, as they are very much sensitive to climatic changes, even in small scale. In rock record, it provides one of the few paleoclimatic indicators in terrestrial environments. A lake is an inland body of water. Lakes form where there is a supply of water to a topographic low on the land surface. Lacustrine (or lakes) environment are diverse; they may be large or small, shallow or deep, and filled with terrigenous, carbonate, or evaporitic sediments. Lakes are fed mainly by rivers and lose water by flow out into a river and/or evaporation from the surface. The balance between inflow and outflow and the rate at which evaporation occurs control the level of water in the lake and the water chemistry. For example, under conditions of high inflow the water level in the lake remain constant, governed by the spill point of the outflow, and the water remains fresh; whereas low water input coupled with high evaporation rates in an enclosed basin results in the concentration of dissolved ions, which may be precipitated as evaporites in a perennial saline lake or when an ephemeral lake dries out. The former variety is grouped under hydrologically open lakes whereas the latter variety forms the hydrologically closed lakes. Lakes can be classified as perennial and ephemeral ones also.

Sedimentation Processes, Stratifications and Circulations within Lakes

Most of the processes that occur in seas also occur in lakes, like waves, currents and tides, but tides are usually incapable of sediment transportation in large scale as lakes are not connected water bodies like seas. Currents in lakes are mainly wave and/or inflow river generated; shore parallel currents are absent, instead two distinct types of river driven currents are common; high density, sediment rich hyperpycnal flows, leading to deposition via turbidity flows, and low density hypopycnal flows, which spread out in upper lake water column. There are, however, important differences with marine settings; like the distinct faunal and floral assemblage, the variation in water chemistry from lake to lake and most importantly, certain physical processes of temperature and density stratification, which are unique to lacustrine environments. Below about 10 or 20 m depth the lake waters are unaffected by any wave or current activity. This allows for the development of lake water stratifications, which developed in response to the differences in temperature, density and/or the chemistry of the waters in the upper and lower parts of the lake. The epilimnion, which is the upper, warmer lake water, are separated from the hypolimnion, the lower, colder part, through a surface across which the temperature changes, the thermocline. The bottom of the lake therefore becomes anaerobic (without air, and therefore oxygen). Seasonal temperature differences may however set up convection circulation in lake waters, leading to free mixing of water. There are a large number of variations in lake circulation and stratification recognized by limnologists (Limnology is the study of lakes). Some of these, applied to lakes deep enough to form a hypolimnion are as below:

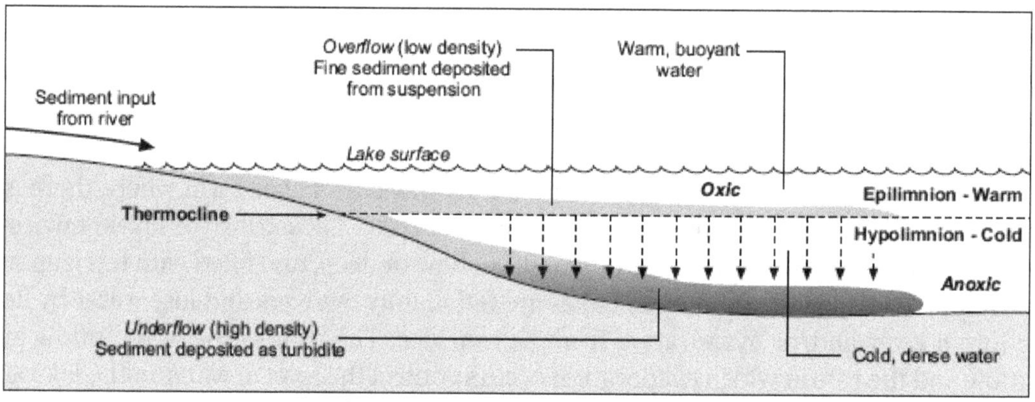

Lake stratifications and current flows.

- Amictic: Lakes permanently isolated from atmosphere by ice cover.
- Cold monomictic: Lake-water never rises above 4 °C, with one period of circulation in the summer.
- Cool diamictic: Lake-water freely circulates twice yearly in spring and autumn.
- Warm monmictic: Lake-water temperatures never drop below 4 °C, freely circulating in winter and stratifying directly in the summer.
- Oligomictic: Lake-water rarely circulates, temperatures always well above 4 °C (i.e., tropical), and stable stratification with small temperature vs. depth variations.

- Polymictic: Lakes with frequent or continuous circulation due to strong winds and/or strong short-term temperature variations.
- Meromictic: Lakes with a chemocline separating a near-permanent bottomlayer that mixes with the main water mass.

Characterization of the Lacustrine Environment

Sedimentation pattern within lakes are primarily controlled by water and sediment discharge amount. These factors may vary widely between lakes within a single climatic zone, hence lacustrine sedimentation pattern records climatic variations adequately. Sedimentation pattern also depends to a great extent whether the lake is hydrologically closed or open. Composition of lake water is also an important factor. Some lakes are of freshwater, others alkaline. A freshwater lake may turn alkaline through time depending on hydrological condition that depends largely on climate and tectonics. Sedimentation pattern also depends on whether a lake is stratified or not and if stratified, how frequently the stratification is disturbed. Siliciclastic, chemical and biogenic sediments are all common in lake environments.

Clastic Lakes

Most siliciclastic sediments are transported into lakes by rivers, although wind-blown, ice-rafted and volcanic material may also be present. Nature and size of drainage basin surrounding lake is a major control on sediment input. Supply of sediment is often seasonally controlled, and hence interlaminated clastic and/or chemical sediments, possibly on a seasonal basis, are quite common. Lacustrine clastic sediments are often thinly bedded and moderately well sorted. Lacustrine subenvironments include deltas, formed where rivers enter the lake, beaches along the margins and deeper offshores.

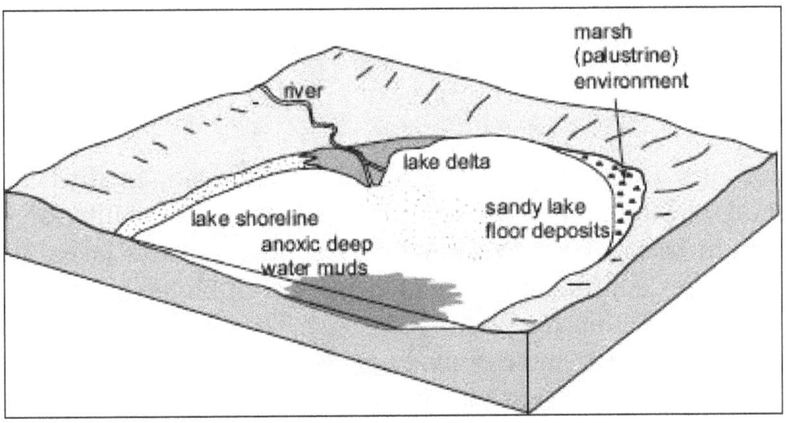

Different subenvironments within clastic lakes.

1. Deltas in Lakes: Where a sediment-laden river enters a lake the water velocity drops abruptly and a delta forms as coarse material is deposited at the river mouth. The form and processes on a lake delta will be similar to that of river-dominated deltas, with

some wave reworking in lakes experiencing strong winds. The character of the delta deposits will be largely controlled by the nature of the sediment supply from the river, and may range from fine-grained deposits to coarse, gravelly fan-deltas. Rapid lake level falls can result in rapid basinward delta progradation.

2. Nearshore areas: Siliciclastic sediments at lake margins concentrated around fluvial entry points. Beaches are narrow because shallow wave base. Abandoned beach ridges may be preserved as a series of staircase ridges that record progressive falls in lake level e.g. Dead Sea. Lake margins may be characterized by sand flats- dry for much of the time, only occasionally flooded during high lake levels.

3. Sheet flood facies are common across low relief sand flats. Nearshore shelves in lake are characteristically sandy with common wave and wave-generated current features. Hummocks and swells are also present in lakes experiencing larger wind fetch. Tidal features are absent though. Desiccation cracks and salt crusts indicate evaporation that exceeds precipitation. Bioturbation may be common.

4. Offshore fines: Most sediment supplied to offshore lake zones by pelagic fallout and turbidity currents. Accumulation rates may be very slow (0.5 mm/yr.). Distribution of fine grained sediments controlled by lake circulation patterns. Slowly accumulating sediments form laminae that may vary on a seasonal basis. Rhythmic or cyclic sediment accumulation takes place. Individual turbidite events, on the other hand, may represent catastrophic fluvial floods. Mixtures of sediment and water brought in by a river or reworked from a lake delta may flow as a turbidity current, which can travel across the lake floor. The deposits will be layers of sediment that grade from coarse material deposited from the current first to finer sediment that settles out last. In lakes where sediment plumes and turbidity currents are the main transport mechanisms the deep lake facies will consist of very finely laminated muds deposited from suspension alternating with thin graded turbidites forming a characteristic, thinly bedded succession.

Carbonate Lakes

Carbonates can form a significant proportion of the succession in any lake setting only if the terrigenous clastic input is reduced. Direct chemical precipitation of carbonate minerals occurs in lakes with raised salinity, but in freshwater lakes, the formation of calcium carbonates is predominantly associated with biological activity. Such precipitation is restricted within the photic zone only. The hard shells of animals such as bivalve molluscs, gastropods and ostracods can contribute some material to lake sediments, and this coarse skeletal material may be deposited in shallow water or redistributed around the lake by wave-driven currents. However, the most abundant carbonate material in lakes is usually derived from algal and microbial sources. While carbonates are the dominant biochemical precipitates, siliceous precipitation also takes place from diatoms and sponges.

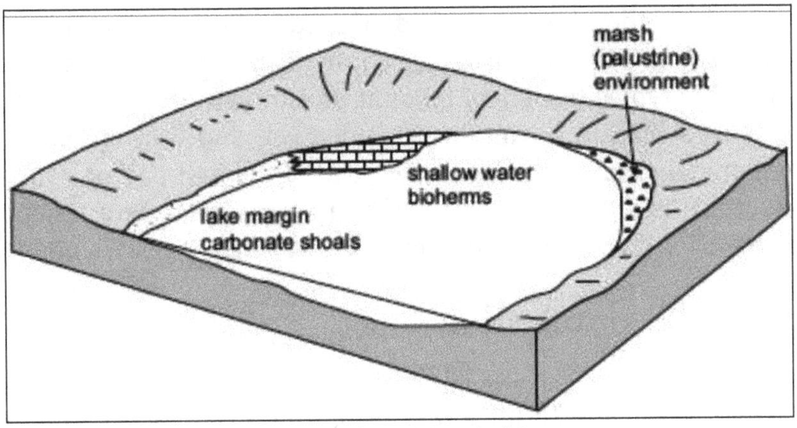

Carbonate depositions in lakes.

Ephemeral Lakes

Ephemeral lakes are large bodies of water that periodically dry out, although the term playa lake is also in use. Once the lake has formed, particles suspended in the water will start to deposit and form a layer of finegrained muddy sediment. Evaporation of the water body gradually reduces its volume and the area of the lake starts to shrink, leaving areas of margin exposed where desiccation cracks may form in the mud as it dries out. With further evaporation the ion concentration in the water starts to increase to the point where precipitation of saline minerals occurs. When an ephemeral lake receives an influx of water and sediment, mud is deposited from suspension to form a thin bed that is overlain by evaporite minerals as the water evaporates. Repetitions of this process create a series of couplets of mudstone and evaporite.

Storms and Shallow Marine Deposits

Most sediments eventually get transported into the oceans. In the sedimentary record, shallow marine deposits are the most abundant due to their large volume and the low erosion rates in those environments. Marine deposits above sea level are often uplifted by tectonics which exposes them to significant erosion. Shallow marine sedimentary deposits include storms, waves, and tides.

Storms

Storms produce both large waves as well as strong and irregular currents. The combination of waves with currents produces unique deposits that can be used to recognize ancient storms in a marine sequence.

Storms generally begin far from the shore and become stronger at the shoreline over time, then either die out or move on. Therefore, near shorelines storms tend to start with low energy flows and increase in speed until the first part of the storm is eroded away, then decrease back to lower energy flows. As the storm wanes, the coarsest sediment is deposited, then finer sand is deposited by waves and currents, and then wave ripples, and so on. The reverse of the sedimentary structures is seen as the storm approaches. There is typically a small amount of sediment deposited at the start of a storm due to some sediment having been eroded yet, and flow speeds increase over time. Consequently, there is typically no recorded of the first half of a storm sequence in the rock record. It is only the second half that is preserved. Storm deposits usually consist of an erosion surface with sediment of decreasing grain sizes upwards.

Hummocky Cross Stratification

Hummocky and swaley cross stratification are deposited as a combination of strong currents as well as large waves unique to storms, and is only found in medium to fine sands. When currents wash eroded sand into an area with strong oscillatory flow from waves, rounded hummocks of sand accumulate on the sea floor separated by lows (swales). These rises are a few to 10 cm high and tens of cm across. Variations in current strength cause local erosion, and the locations of hummocks and swales changes over time. This produces erosional surfaces that truncate older laminae. HCS is characterized by low angle laminae truncated by low angle surfaces. There are many concave and convex up laminae and fewer flat laminae.

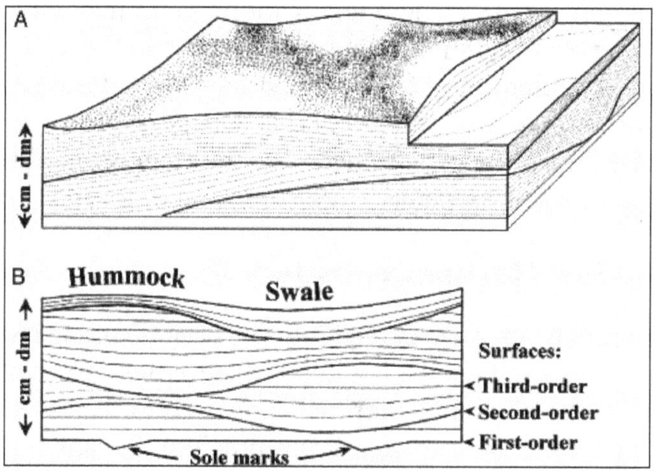

HCS Storm Sequence.

Waves

It is important to know that waves have bi-directional flow every few seconds. The oscillating current in both directions is the same in deep water, however this is not always the case near the shore.

Wave Ripples

- Wave ripples experience transport in both directions every few seconds, unlike current ripples. In rocks, they have a distinctive symmetric shape when the flow in each direction is the same speed. Additionally, cross-laminations that dip in two directions are also present, regardless of the ripples being symmetric.

- At low flow ripple crests are sharp: This is because the boundary layer does not have enough speed or momentum to get rid of the crest of the ripple and the grains that are moved are deposited on the upper part of the slope.
- High flow ripples have rounded crests: This is due to the crests eroding and deposition occurring further down the slope.

- In shallow water, currents from waves can be strong enough to flatten ripples. Since the flow switches directions too frequently, dunes are unable to build up. Therefore, where flow speeds are too fast to form ripples, the sedimentary surface tends to be planar or broadly scalloped as the waves are focused into certain areas. This produces flat lamination (different from upper planar lamination), where waves are in very shallow water relative to their height.

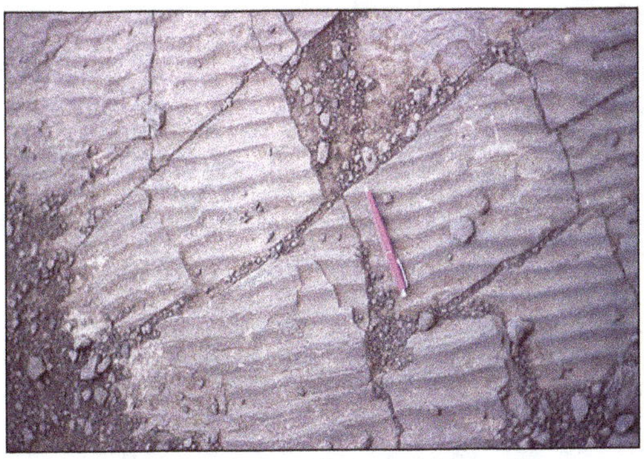

Wave ripples, with a pen for scale.

Tides

Two key characteristics that are unique to tides are that flow changes direction once or twice per day, and the speed of flow is cyclical. The flow goes onshore, stops at high tide, then flows offshore and stops at low tide. There is a great deal of variability in tides depending on geography. Flow speeds vary, consequently producing different sedimentary structures. The main characteristic to look for is the variations in flow speed and direction. Another important point about tidal processes is that bi-directional flow can have varying speeds over hours.

- Tidal sedimentary structures:
  - Herringbone cross stratification.

Reactivation Surfaces

Reactivation surfaces form when flow in one direction is stronger than the other, but the other flow is still strong enough to modify the bedform shape. Reactivation surfaces are erosion surfaces within the sets of cross stratification. They looks like irregular surfaces that are oriented similarly to the foresets, but usually have a shallower dip. Additionally, the foresets above and below the reactivation surface typically have slightly different orientations. As is common in tidal environments, reactivation surfaces indicate varying flow directions.

Mud Drapes

Mud drapes are one type of sedimentary structure that can form from cyclical flow speeds. During slack tides (low or especially high), fine grained sediment can fall out of suspension draping tidal bedforms with mud. Since mud is cohesive, it does not always erode during the following tidal flow, particularly in the separation zone where there is a slow flow. Therefore, sand foresets coated with mud are very common in tidal environments.

Mud drapes outcrop.

References

- Sedimentology-and-stratigraphy, geology-and-petroleum, petroleum-geology: aapg.org, Retrieved 21, July 2020
- A-Importance-of-Sediment, A-Draft-Textbook, A-Sediments-and-Strata-(Sumner), University-of-California-Davis: libretexts.org, Retrieved 14, February 2020
- A-Sedimentary-Structures, A-Weathering-Erosion-and-Sedimentary-Rocks, An-Introduction-to-Geology-(Johnson-Affolter-Inkenbrandt-and-Mosher), Geology: libretexts.org, Retrieved 26, May 2020
- A-Glacial-Environments, Glacial-environments-are-defined-as-occurs-when-the-ice-melts, A-Sediments-and-Strata-(Sumner), A-Old-or-Lost-Pages, A-Alluvial-Fans, University-of-California-Davis: libretexts.org, Retrieved 15, January 2020
- Storms-and-Shallow-Marine-Deposits, A-Sediments-and-Strata-(Sumner), University-of-California-Davis: libretexts.org, Retrieved 06, April 2020

Permissions

All chapters in this book are published with permission under the Creative Commons Attribution Share Alike License or equivalent. Every chapter published in this book has been scrutinized by our experts. Their significance has been extensively debated. The topics covered herein carry significant information for a comprehensive understanding. They may even be implemented as practical applications or may be referred to as a beginning point for further studies.

We would like to thank the editorial team for lending their expertise to make the book truly unique. They have played a crucial role in the development of this book. Without their invaluable contributions this book wouldn't have been possible. They have made vital efforts to compile up to date information on the varied aspects of this subject to make this book a valuable addition to the collection of many professionals and students.

This book was conceptualized with the vision of imparting up-to-date and integrated information in this field. To ensure the same, a matchless editorial board was set up. Every individual on the board went through rigorous rounds of assessment to prove their worth. After which they invested a large part of their time researching and compiling the most relevant data for our readers.

The editorial board has been involved in producing this book since its inception. They have spent rigorous hours researching and exploring the diverse topics which have resulted in the successful publishing of this book. They have passed on their knowledge of decades through this book. To expedite this challenging task, the publisher supported the team at every step. A small team of assistant editors was also appointed to further simplify the editing procedure and attain best results for the readers.

Apart from the editorial board, the designing team has also invested a significant amount of their time in understanding the subject and creating the most relevant covers. They scrutinized every image to scout for the most suitable representation of the subject and create an appropriate cover for the book.

The publishing team has been an ardent support to the editorial, designing and production team. Their endless efforts to recruit the best for this project, has resulted in the accomplishment of this book. They are a veteran in the field of academics and their pool of knowledge is as vast as their experience in printing. Their expertise and guidance has proved useful at every step. Their uncompromising quality standards have made this book an exceptional effort. Their encouragement from time to time has been an inspiration for everyone.

The publisher and the editorial board hope that this book will prove to be a valuable piece of knowledge for students, practitioners and scholars across the globe.

Index

www.ingramcontent.com/pod-product-compliance
Lightning Source LLC
LaVergne TN
LVHW080805220926
848485LV00011B/205
* 9 7 8 1 6 3 9 8 7 5 2 0 7 *